Armies of the US -Mexican War, 1846-1848

By
Gabriele Esposito

Color plates by
Benedetto Esposito

The Armies of the Mexican-American War, 1846-1848
By Gabriele Esposito
Color plates by Benedetto Esposito
Cover by Huge Charles McBarron, Jr. (Department of the Army)

This edition published in 2020

Winged Hussar is an imprint of

Winged Hussar Publishing, LLC
1525 Hulse Rd, Unit 1
Point Pleasant, NJ 08742

ISBN 978-1-950423-40-8
LCN 2020938179

Bibliographical References and Index
1. Military History. 2. Mexican-American War. 3. 19th Century American

For more information
visit us at www.wingedhussarpublishing.com

Twitter: WingHusPubLLC
Facebook: Winged Hussar Publishing LLC

TABLE OF CONTENTS

Preface

This book is dedicated to my parents, Maria Rosaria and Benedetto, for their great love and precious support during the creative process of the present work. A very special thanks is due to my father, for creating the magnificent uniform plates that illustrate the present work: his love for colour and his passion for art have transformed some uniforms into "living" subjects. Special thanks are due to the editor of this volume, Vincent Rospond, for having enjoyed and sponsored this project since the early beginnings. Very special thanks goes to Olivier Millet, a great researcher of the US-Mexican War, for giving me permission to use the fantastic uniform plates that he has created on the armies that fought during 1846-1848. The colour plates reproduced in this book and realised by Olivier are based on original models by Alexis Cabaret, which are available on the website: http://centjours.mont-saint-jean.com/; Olivier Millet, instead, has a fantastic blog which displays most of his personal researches and colour plates: http://history-uniforms.over-blog.com/. Another friend who gave an enormous contribution to the present work is Patrick Berthelot, an excellent military historian and illustrator living in France who has realised all the dozens of uniform plates dedicated to the Mexican Army that are reproduced in this book. These are extremely detailed and cover each single unit of Santa Anna's military forces. They have been produced after years of research and work: this makes them something absolutely special. Thanks to Patrick's generosity, it has been possible to publish them in this book in order to give an idea of how complex and rich is the Mexican uniformology.

The main aim of this book is to present a detailed overview of the military forces involved into the US-Mexican War of 1846-1848; most of the text will be devoted to the organization of the opposing armies, which included troops of various kinds. For the US forces the analysis will focus on the years of the conflict, while for the Mexican Army the analysis will start from Mexico's independence in 1821. This organization of the text is the result of a specific choice from the author: understanding the military organization of the Mexican Army during the war of 1846-1848, in fact, is not possible without knowing which units made up the Mexican military forces during the delicate period of 1821-1844. As a result of the above, the text will have the following structure on four chapters: the first one will deal with the history and organization of the US Army from the early mobilization of 1845 to the end of the hostilities with Mexico in 1848; the second one will cover the Mexican military forces during the crucial years of 1821-1848; the third chapter will be devoted to the military uniforms of the USA (including the little-known volunteer corps); the fourth chapter will be dedicated to the uniforms of all the Mexican military forces that took part to the conflict. The two chapters on armies will have specific sections devoted to the weapons employed by the soldiers of both sides.

Gabriele Esposito

Chronology

March 31, 1845: Mexico breaks diplomatic relations with the USA

June 15, 1845: General Zachary Taylor moves with the US Army to Texas

July 4, 1845: The Republic of Texas joins the United States of America

January 13, 1846: The US Army moves to the southern border of Texas

April 25, 1846: First skirmishes on the border between American and Mexican troops

May 8, 1846: Battle of Palo Alto

May 9, 1846: Battle of Resaca de la Palma

May 13, 1846: The USA declares war on Mexico

June 5, 1846: The US "Army of the West" starts its march from Kansas to Santa Fe

July 4, 1846: California becomes independent as the "Bear Flag Republic"

July 9, 1846: The US Navy occupies San Francisco in California

July 19, 1846: The "Bear Flag Republic" comes under control of the US government

July 29, 1846: The US Navy occupies San Diego in California

August 13, 1846: The US Navy occupies Los Angeles in California

August 16, 1846: General Santa Anna returns to Mexico after a brief exile in Cuba

August 17, 1846: The US "Army of the West" occupies Santa Fe in Mexico

September 20, 1846: Battle of Monterrey

September 25, 1846: The US "Army of the West" starts its march towards California

November 18, 1846: Major General Winfield Scott is appointed to command an expedition directed against Vera Cruz, in Mexico

November 23, 1846: Scott's expeditionary force leaves for Mexico

December 6, 1846: Santa Anna resumes his complete control over Mexican government

December 12, 1846: The US "Army of the West" reaches San Diego in California

January 13, 1847: The whole territory of California is under US control

February 22-23, 1847: Decisive Battle of Buena Vista between Taylor and Santa Anna

March 9, 1847: Scott's expeditionary force lands at Vera Cruz

March 29, 1847: Vera Cruz surrenders to the US military forces

April 2, 1847: Scott's army prepares to move on Mexico City

April 18, 1847: Decisive Battle of Cerro Gordo between Scott and Santa Anna

May 4, 1847: Most of the US volunteers return back home

August 19-20, 1847: Battle of Churubusco

September 8, 1847: Battle of Molino del Rey

September 13, 1847: Battle of Chapultepec

September 14, 1847: The US Army occupies Mexico City

September 16, 1847: Santa Anna resigns as "president" of Mexico

February 2, 1848: The Treaty of Guadalupe-Hidalgo is signed, ending the war. The USA obtains the following territories from Mexico: California, Nevada, Utah, Arizona, Colorado, New Mexico and Texas.

The military campaigns of the US-Mexican War

Map showing the new boundaries fixed at the end of the war

James Knox Polk (left) and General Santa Anna (right), the leaders of the USA and Mexico during the war

The US Army of 1845-1848

During the first half of the 19th century, the US Army remained a very modest institution if compared with the contemporary European armies. It comprised only a small number of regular soldiers, who were scattered across the many garrisons of the country and who had as main function that of protecting the colonists' settlements of the western frontier from native attacks. After the disbandment of George Washington "Continental Army", the USA had always deployed small regular forces that were supplemented by the much more numerous militias of the single states. It was only during the War of 1812 against Great Britain, albeit temporarily, that the US Army was significantly enlarged in order to face the needs of a large-scale conflict. The scarce consistency of the US regular forces was due to a series of reasons; first of all, the borders of the nation were not menaced by any foreign power. On the northern border, relations with British Canada remained quite positive after 1815; on the southern border, Mexico had serious difficulties in keeping possession of the vast scarcely-populated regions stretching from Texas to California. As a result, the only concrete enemy of the USA were the hostile native tribes and these could be easily controlled by a small military force of regulars. In addition to the above, it is important to remember that the American government and public opinion had always had little trust in the regular army. The latter was perceived as a potential menace to the stability of the democratic system, since it could influence the political life of the country as it happened in all the young republics of Latin America (where presidents were chosen by the armed forces and military coups were extremely frequent). Most of the Americans considered the state militias as the only true "democratic" organization that could defend the republic from foreign menaces; any large regular force would have been a political instrument in the hands of the ruling classes that could have been used to impose the will of a minority.

A Colonel of the US Army in 1846

As the events of the War of 1812 had already shown, however, a large nation having the ambition of dominating over the Americas absolutely needed a sizeable permanent force. The state militias, being formed by non-professional soldiers, had little training and could not be employed for long periods of time outside the borders of their home state. This meant that they could be used for defensive operations, but that in case of an expansionist war the government would have been obliged to expand the regular army by recruiting large numbers of volunteers. Generally speaking, this is exactly what happened after the outbreak of the war with Mexico when the US Army invaded the enemy territory. In 1845, also due to economic concerns regarding the high costs of the army, the regular military forces of the USA did comprise only 7,300 soldiers; these had to protect a nation with a population of nearly 20 million people and a territory of 1.8 million of square miles. The US Army was under control of a civilian secretary who headed the Department of War and did comprise several bureaucratic elements. The latters consisted of 259 staff officers, 17 military storekeepers and 450 enlisted men organized into nine different staff departments. Each of these was headed by a different senior officer according to its main function: Adjutant General, Inspector General, Paymaster General, Quartermaster General, Commissary General of Subsistence, Surgeon General, Chief of Ordnance, Chief of Engineers and Chief of Topographical Engineers. All the staff departments operated in an autonomous way and were mostly charged with all the administrative/logistical aspects of military life. All the officers and soldiers who were part of the nine departments had no combat roles and this furtherly reduced the effective number of regulars who could be employed on the field in case of war. Most of the American soldiers were rude and illiterate individuals who had not been able to find their place in an industrious society like that of the expanding United States; as a result, all the other citizens did not have a very high opinion of them and thus they lived at the margins of society. Under-funded, under-manned and dispersed into small frontier outposts, the US regulars had little opportunities of social advancement. The situation was different for officers, who came from the ruling classes of society and who had a good personal preparation; the latter was given them from the famous United States Military Academy of West Point. This, founded in 1802, was one of the best in the world.

An officer of the US Dragoons in the 1840's

Infantry

At the beginning of the war, in 1845, the infantry of the US Army consisted of eight line regments numbered from 1 to 8. Each regiment was made up of ten companies; eight of the latters were known as "battalion companies" and were made up of fusiliers; the remaining ones were known as "flank companies" and consisted of grenadiers and riflemen. The grenadier company was to be posted on the right flank of the regiment and the rifle one was to be posted on the left flank of the regiment, hence their peculiar denomination. In practice, however, there was very little difference between the companies of fusiliers and the two flanking ones; only weaponry was different, since the fusiliers were armed with smoothbore muskets while the grenadiers/riflemen were equipped with rifled muskets (for skirmishing or sharpshooting). The staff of each infantry regiment did comprise the following elements: one colonel (commanding the unit), one lieutenant-colonel (second in command), one major, one adjutant, one quartermaster, one quartermaster sergeant, one sergeant major and two musicians. The composition of each infantry company, instead, was as follows: one captain, one lieutenant, one second-lieutenant, two musicians, four sergeants, four corporals and 42 soldiers. Five companies, half a regiment, could be assembled to form temporary battalions. Battalions comprising companies from different regiments, even more than five, could be formed to perform specific duties or in case of need. Until 23 August 1842 there had been 90 soldiers in each company, but on that date this establishment was reduced to 42 by the Congress due to economic reasons. During the war, on 13 May 1846, a new act from the Congress increased the number of soldiers in each infantry company to 100; this new establishment, however, was rarely reached.

On 11 February 1847, when it became clear that more regular infantry units were needed for the invasion of Mexico, the Congress authorized the formation of another nine infantry regiments. These would have comprised eight line infantry regiments (thus doubling the number of existing line units) plus one elite "Regiment of Voltigeurs and Foot Riflemen". The new line regiments would have been numbered from 9 to 16, while the elite light infantry unit would have been simply known by its peculiar denomination. The internal organization of the eight new line regiments was exactly the same of the existing ones, except for the addition of another second lieutenant for each company. The "Regiment of Voltigeurs and Foot Riflemen" had a very peculiar structure and tactical function. It was to be organized on five companies of "foot riflemen" (light infantrymen) and five of "voltigeurs" (mounted infantrymen) plus one company/battery of mountain howitzers and rockets. The regiment would have constituted a combined-arms force comprising elite elements of infantry, cavalry and artillery: when rapid movement was needed, each foot rifleman would have climbed up on the horse of a mounted voltigeur; when there was the need for a static formation, the mounted voltigeurs would have dismounted to fight side by side with the foot riflemen. In addition, the ten companies would have been supported by the fire of the mountain artillery company. The general idea behind the formation of the "Regiment of Voltigeurs and Foot Riflemen" was a very innovative one: the unit would have been a fast-moving light corps able to operate very effectively also on the broken terrain of the Mexican mountains. The original plans described above, however, were never put in practice: the regiment, in fact, never received the needed horses for the five compa-

US VOLTIGEURS AND FOOT RIFLEMEN

Sub-lieutenant | Soldier, Corporal – grey uniform | Soldier, Corporal – infantry uniform | Soldier – dark blue dress

nies of voltigeurs and thus had to renounce to its planned high mobility. Until its disbandment at the end of the war with Mexico, the "Regiment of Voltigeurs and Foot Riflemen" continued to operate as a simple infantry unit having some distinctive "light" features (since its members were all equipped with excellent M1841 "Mississippi" rifle, the first US rifled musket having a percussion-lock system).

Cavalry

At the beginning of the war the cavalry of the US Army consisted of just two dragoon regiments, which had acquired a great fighting experience on the frontier while campaigning against the natives. As a result, the dragoons were with no doubts the most battle-hardened veterans of the whole army and the only American soldiers who were used to gruelling marches on broken terrain. Before the war with Mexico, the regiments had never operated as a single entity since their companies were usually scattered on several frontier outposts located in the West. The US dragoons were excellent light cavalrymen and on most occasions they operated as mounted infantrymen, thus using horses only to travel long distances and dismounting to fight. They were not capable of operating like a "shock force" and had very little experience of frontal charges (differently from their Mexican opponents). The internal organization of the two cavalry regiments was very similar to that of the infantry ones; they comprised ten companies each and there were no squadrons. Each single dragoon company included the following elements: one captain, one lieutenant, one second-lieutenant, four sergeants, four corporals, two buglers, one farrier, one blacksmith, two teamsters (civilians) and 50 troopers. The staff of each cavalry regiment comprised the following elements: one colonel (commanding the unit), one lieutenant-colonel, one major, one adjutant, one quartermaster, one sergeant major, one quartermaster sergeant, one chief musician, two chief buglers and one principal teamster (who was a civilian contractor). Every company of each regiment had horses of a different colour, so that it was quite easy to recognize single companies.

US DRAGOONS

Sub-lieutenant Soldier Sergeant Soldier Corporal Soldier

On 13 May 1846 the Congress increased the number of troopers in each dragoon company to 100, but like for the infantry these numbers remained only on paper. A few days later, on 19 May, the Congress also authorized the creation of a new cavalry regiment. This unit, known as "Regiment of Mounted Riflemen", was to be an elite corps of mounted light infantry; its main purpose was that of guarding and patrolling the long "Oregon Trail" that connected Independence in Missouri with Oregon City on the Pacific Ocean. Due to the war needs, however, also the "Regiment of Mounted Riflemen" was finally employed on the Mexican theatre of operations. This unit had exactly the same internal organization of the two existing units of dragoons, but it had a distinct "light" character (like its foot equivalent, the "Regiment of Voltigeurs and Foot Riflemen", it was equipped with the new M1841 "Mississippi" rifle). On 11 February 1847, in order to have more mounted troops for the invasion of Mexico, the Congress decided to authorize the creation of a 3rd Dragoon Regiment; this was to have the same establishment of the existing regiments but with the addition of one surgeon and two assistant surgeons. At the end of the war with Mexico, this new regiment was disbanded and thus the US Army retained only three units (two regiments of dragoons and one of mounted riflemen). Generally speaking, the US cavalry performed extremely well during the conflict with Mexico but on several occasions it suffered heavy casualties due to the great quality of its Mexican opponents (especially the irregular contingents of "vaqueros", armed with lances and "lazos").

Artillery and Technical Corps

The artillery was with no doubts the most professional branch of service of the US Army; it was equipped with some of the best guns in the world and many of its officers were extremely competent because they had graduated at West Point. In 1824 the US Army had also created a specific Artillery School at Fortress Montroe, in Virginia: this worked as an advanced school for artillery officers and as a training ground for artillery units to practice their fighting skills. In 1840 the US government had sent a board of officers to Europe with the objective of studying the latest developments in artillery weapons. This military mission returned to Washington after having visited several European countries and brought to the United States samples of various types of artillery pieces produced by the European industries. The information and experience obtained thanks to this military mission sent to Europe were fundamental for the US artillery, since the American manufacturers of artillery pieces could analyse the latest models of European guns and thus improve/update their own products; in addition the board that had been sent to Europe designed a new and complete artillery weapons system that was codified in the "Ordnance Manual for the Use of the Officers of the United States Army" of 1841. Some years

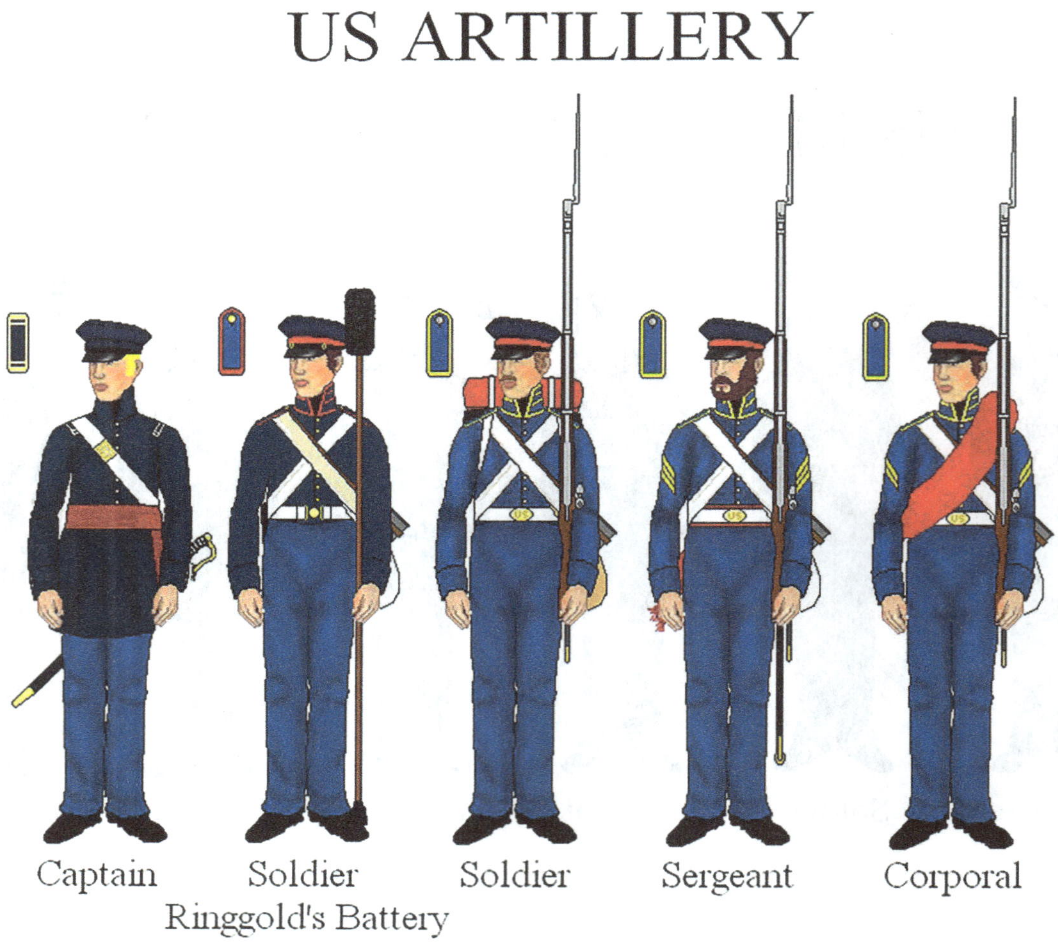

later, in 1845, this was followed by the "Instruction for Field Artillery, Horse and Foot", which was based on French and English artillery instructions. During the years up to 1839, the artillery companies of the US Army mostly served as independent units in the wars against the natives and were scattered on several frontier outposts; on most occasions, due to the same nature of the military operations against the natives, the artillerymen were employed as simple infantrymen and thus had little opportunity to man their field pieces.

To solve this problem, in 1839, a special training camp was established at Camp Washington in New Jersey. Here all the American artillerymen were re-trained during a period of intensive activities and learned all the latest updates introduced in the field of artillery tactics. After the few months spent at Camp Washington, the artillerymen of the US Army became the best ones of the whole American continent. At the beginning of the hostilities with Mexico, the US artillery comprised four regiments numbered from 1 to 4. Each regiment had ten companies/batteries, lettered from A to K (there was no "Company J"). According to the artillery reorganization of 1821, one company/battery in each regiment was to be equipped with horses and light pieces in order to act as mounted/light artillery; in practice, however, it was not until 1838 that this measure was put into practice for the first time in the 3rd Artillery Regiment. Major Samuel Ringgold, the greatest light artillery expert of the US Army, organized the first mounted company/battery; this soon achieved a superior level of training and proficiency that greatly encouraged the rapid expansion of Ringgold's program. By 1844 the following companies/batteries of the US artillery had been transformed into mounted/light ones by Major Ringgold: Company K of the 1st Artillery Regiment, Company A of the 2nd Artillery Regiment, Company C of the 3rd Artillery Regiment and Company D of the 4th Artillery Regiment. Shortly before the outbreak of the hostilities with Mexico, another company was re-trained as mounted/light artillery; this was Company E of the 3rd Artillery Regiment (Ringgold's unit). During the war the companies of mounted/light artillery were usually detached from their regiments and thus operated as independent units. This was the standard internal structure of an artillery company/battery: one captain (commanding the unit), two lieutenants, one second-lieutenant, four sergeants, four corporals, two musicians, two artificers and 54 gunners. Until 1842, the gunners of each company had been 71 and there had been also a third artificer. Later, when Congress authorized the increase of infantry and cavalry companies up to 100 men, the same was made also for the artillery; as for the infantry and cavalry, however, the new establishments were reached quite rarely. The

US Artillery officer in 1846

staff of an artillery regiment comprised the following elements: one colonel (commanding the unit), one lieutenant-colonel, one major, one sergeant major and one quartermaster sergeant. In time of peace each company/battery was to have four guns, with another two that were put into storage; in case of war, also the two "reserve" guns would have been activated thus bringing the total number of pieces in a regiment to 60 guns. During the war, on 3 March 1847, Congress authorized the formation of a second mounted/light company in each regiment; only two of the existing foot companies, however, were converted to mounted/light ones: Company I of the 1st Artillery Regiment and Company G of the 4th Artillery Regiment. The war period also saw the formation of a new "experimental" unit, known as "Howitzer and Rocket Company". This was manned by personnel from the Ordnance Department (former non-combatants) and was equipped with mountain howitzers and rockets. This special artillery company was later attached to the "Regiment of Voltigeurs and Foot Riflemen", as described before.

Like in the Mexican Army, the US armed forces comprised a "Corps of Engineers" (made up of specialist officers) and a separate "Company of Sappers, Miners and Pontoniers". The members of the latter unit, which was organized only after the outbreak of the war with Mexico on 15 May 1846, acted as the enlisted soldiers of the Corps of Engineers. The latter comprised just a few officers who had specific "technical" skills and who were based at the Military Academy of West Point; these had more or less the same functions of their Mexican equivalents and were among the best officers of the whole US Army. The new unit of sappers/miners/pontoniers had the consistency of a double-sized company with 100 men: ten sergeants, ten corporals, two musicians, 39 artificers and 39 laborers. Those with the qualification of "sappers" were required to act as combat engineers; the miners were specialized in mining and counter-mining operations; pontoniers were required to handle and construct pontoon bridges.

In addition to the above, the US Army did comprise also another very "special" unit having specific purposes: the so-called "Mexican Spy Company". This was a cavalry company made up of Mexican renegades who wished to serve under the American flag; these were mostly ex-bandits who had been freed from the Mexican prisons due to the war or deserters from the Mexican Army. They were under command of Manuel Dominguez, one of the first Mexicans who had joined the US Army. With the progression of time, this "corps" was expanded to two companies with 100 men each; armed with lances and dressed with Mexican civilian dress, the members of the Spy Company were under direct command of General Winfield Scott (commander-in-chief of the US Army in Mexico) and were paid $20.00 a month for their "services". The Mexicans of this special company acted as couriers, escorts, guides, interpreters, scouts and spies for the US Army; as a result, they proved to be extremely precious for the American cause and performed many of the duties that would have been normally assigned to the regular cavalry. The soldiers of the "Spy Company" were obviously hated by all the Mexicans who had remained loyal to their motherland, especially because they were extremely efficient in obtaining information about Santa Anna's plans and operations. In addition, these "renegade" soldiers fought with great valour against the Mexican irregular "guerrilleros" by using their same methods of warfare. The "Spy Company" was disbanded in June 1848 and after the end of the war most of its members went to the United States.

Finally, to complete this general description of the American "special" corps, it is necessary to spend a few words on the US Corps of Marines. This was probably the best unit of the whole American military, mostly thanks to its superior training and excellent equipment. The Marines, who were part of the Navy and not of the Army, were naval infantrymen capable of

US MARINES

Captain Soldier parade dress Corporal Sergeant Soldier

serving aboard warships but also of forming "landing parties" for amphibious operations. In 1845, at the outbreak of the war, the Marine Corps comprised 42 officers and 986 enlisted men; these were employed on several occasions during the conflict mostly to secure captured coastal areas and to fight on board of warships. In May 1847 the overall commander of the Marine Corps, Brigadier-General Archibald Henderson, proposed that a temporary "Regiment of Marines" could be raised and sent to fight in Mexico as an elite infantry force. According to Henderson's view, this new Marine unit could have been raised more rapidly and more economically than a regiment of line infantry; his plan was accepted, but the recruiting operations for the new unit proved to be quite difficult. According to the original plans, the "Regiment of Marines" would have comprised a total of 1,000 enlisted men; in practice, however, it was possible to raise only six companies for service in Mexico. These were organized into two battalions, one of which was finally sent to the theatre of operations. Here the young recruits of the new "Regiment of Marines" fought with great valour during the last phases of the campaign, distinguishing themselves in the "Halls of the Montezumas".

Volunteers

On 11 May 1846, the same day President Polk informed Congress that a state of war existed between the USA and Mexico, a bill was passed by both the House of Representatives and the Senate which authorized the use of the state militias and the recruiting of 50,000 volunteers. According to US federal law, all able-bodied men aged between 18 and 45 were required to enrol in the militia of their home state. This meant that each state, according to its population, had a different number of militia corps. The standard militia unit was the company, which comprised the following elements: two officers, twelve NCOs and 64 men. Ten companies usually made up a militia regiment. Militia corps did comprise infantry and cavalry units, but also some little artillery ones. In theory, according to the current population of the nation, in 1846 the US government could potentially mobilize little less than two millions of militiamen; in practice, however, most of the militia units existed only on paper and in any case they could be deployed only for three months of service. Due to the fact that the entire war was fought on Mexican soil, the state militias could never be employed to support the regular army: according to US law, in fact, the militia was a home defence force and thus could be deployed only on the American territory. As a result, only the volunteers could be employed in Mexico and their contribution was absolutely fundamental for the final victory of the USA.

Illinois State Battle Flag from the US-Mexican War

As anticipated above, the US regular army was too small to fight in a large-scale war; as a result, the US government absolutely needed massive numbers of volunteers in order to increase the general consistency of its military forces. According to the bill of 11 May 1846, each volunteer was to serve for a period of one year; after this, he could decide to continue serving with the army for the rest of the war or return home. The volunteers were to furnish their own uniforms and personal equipment, except for the weapons that were provided by the central government. All volunteers received the same pay of the regulars and were compensated by the central government in case their uniforms or pieces of equipment were damaged in combat. The senior officers of the volunteer units were to be appointed according to the laws of their home state; some of them were quite good, especially those who had graduated at West Point, while others had very little military experience and thus did not perform well during the war. Junior officers were usually elected by their own men and were often quite incompetent. Generally speaking, the single volunteers were good potential soldiers who were accustomed to harsh conditions of life but had little military discipline to speak of. When the conflict broke out, war fever was already running high across all the states of the nation: thousands and thousands of men responded to the call for volunteers and enlisted in the ranks of the new units that were being formed in each state. The first half of the volunteers was recruited from the states of Alabama, Arkansas, Georgia, Illinois, Indiana, Kentucky, Mississippi, Missouri, Ohio, Tennessee and Texas; the second half, which was to be deployed during the later course of the war, was recruited from the remaining states (California, District of Columbia, Florida, Iowa, Louisiana, Maryland, Massachusetts, Michigan, New Jersey, New York, North Carolina, Pennsylvania, South Carolina and Virginia). In many places the quotas of volunteers were filled in a matter of hours and many states had much more men that those effectively needed: the patriotism of the American people was at its highest level and everyone wanted to give his personal contribution to the progress of the nation. The idea of "manifest destiny" was growing stronger and all Americans wished to expand the borders of their country by conquering the rich lands of northern Mexico. Due to the great enthusiasm of the population, most of the volunteer companies were overstrength and had 100 men instead of the authorized 64; they were accepted on a first come basis, since there were too many men wishing to enlist. Thousands of volunteers had to be left home, simply because they were not needed and there was no space for them. With the progression of time, when it became apparent that the war with Mexico would have been much more difficult and longer than expected, the enthusiasm of the volunteers started to vanish. Desertion and diseases began to thin the ranks of the volunteer units, while most of the latters ended their short-term enlistment and returned back home. Now we will report, for each state, all the volunteer units that were formed for the war against Mexico.

Alabama

- Battalion of Alabama Volunteers, on 4 companies, May-August 1846
- Battalion of Alabama Volunteers (new), on 5 companies, November 1847-June 1848
- 1st Independent Company of Alabama Volunteers, May-August 1846
- 2nd Independent Company of Alabama Volunteers, May-August 1846
- 3rd Independent Company of Alabama Volunteers, May-August 1846
- Independent Company of Alabama Mounted Volunteers, June 1847-July 1848

Militiamen from New Hampshire in 1845

Arkansas

- Regiment of Arkansas Mounted Volunteers, on 10 companies, June 1846-June 1847
- 1st Independent Company of Arkansas Mounted Volunteers, May 1847-June 1848
- 2nd Independent Company of Arkansas Mounted Volunteers, June 1847-June 1848

California

- Battalion of California Volunteers, on 9 companies (eight of mounted riflemen and one of artillery), July 1846-April 1847

Florida

- 1st Independent Company of Florida Volunteers, February 1847-March 1848
- 2nd Independent Company of Florida Volunteers, August 1847-July 1848

Georgia

- Regiment of Georgia Volunteers, on 10 companies, June 1846-May 1847
- Battalion of Georgia Volunteers, on 5 companies, July 1847-July 1848
- Battalion of Georgia Mounted Volunteers, on 6 companies, September 1847-July 1848
- Independent Company of Georgia Mounted Men, May 1847-August 1848

Illinois

- 1st Regiment of Illinois Volunteers, on 10 companies, June 1846-June 1847
- 2nd Regiment of Illinois Volunteers, on 10 companies, June 1846-June 1847
- 3rd Regiment of Illinois Volunteers, on 10 companies, July 1846-May 1847
- 4th Regiment of Illinois Volunteers, on 10 companies, July 1846-May 1847
- 1st Regiment of Illinois Volunteers (new), on 10 companies, June 1847-October 1848
- 2nd Regiment of Illinois Volunteers (new), on 10 companies, August 1847-July 1848
- 1st Company of Illinois Mounted Volunteers, May 1847-November 1848
- 2nd Company of Illinois Mounted Volunteers, August 1847-July 1848
- 3rd Company of Illinois Mounted Volunteers, August 1847-October 1848
- 4th Company of Illinois Mounted Volunteers, September 1847-July 1848

Indiana

- 1st Regiment of Indiana Volunteers, on 10 companies, June 1846-June 1847
- 2nd Regiment of Indiana Volunteers, on 10 companies, June 1846-June 1847
- 3rd Regiment of Indiana Volunteers, on 10 companies, June 1846-June 1847
- 4th Regiment of Indiana Volunteers, on 10 companies, June 1847-July 1848
- 5th Regiment of Indiana Volunteers, on 10 companies, October 1847-July 1848

INDIANA VOLUNTEERS

Captain Soldier Sergeant Soldier Standard-bearer

Iowa

- Battalion of Iowa "Mormon" Volunteers, on 5 companies, July 1846-July 1847

Kentucky

- Regiment of Kentucky Cavalry Volunteers, on 10 companies, June 1846-July 1847
- 1st Regiment of Kentucky Volunteers, on 10 companies, May 1846-May 1847
- 2nd Regiment of Kentucky Volunteers, on 10 companies, June 1846-June 1847
- 3rd Regiment of Kentucky Volunteers, on 10 companies, October 1847-July 1848
- 4th Regiment of Kentucky Volunteers, on 10 companies, October 1847-July 1848
- Independent Company of Kentucky Volunteers, May 1846-May 1847

Louisiana

- 1st Battalion of Louisiana Volunteer Artillery, on 2 companies, August 1845-November 1845
- 2nd Battalion of Louisiana Volunteer Artillery, on 3 companies, May 1846-July 1846
- Brigade of Louisiana Volunteers, on 6 regiments with 10 companies each, May-August 1846
- Independent Company of Louisiana Volunteers, July 1846-May 1847
- Regiment of Lousiana Volunteers, on 10 companies, December 1846-July 1848
- Battalion of Louisiana Volunteers, on 5 companies, May 1847-July 1848
- Battalion of Louisiana Mounted Volunteers, on 5 companies, July 1847-July 1848

Maryland and District of Columbia

- Battalion of Maryland and District of Columbia Volunteers, on 6 companies (five of infantry and one of artillery), May 1846-May 1847
- Regiment of Maryland and District of Columbia Volunteers, on 8 companies, May 1847-July 1848
- Tennessee Company of Mounted Volunteers (attached to the Maryland Regiment), May 1847-July 1848
- Pittsburg Company, October 1847-July 1848
- Independent Company of Maryland Volunteers, June 1847-August 1848

Massachusetts

- Regiment of Massachusetts Volunteers, on 10 companies, January 1847-July 1848

Michigan

- Regiment of Michigan Volunteers, on 10 companies, December 1847-July 1848
- Independent Company of Michigan Volunteers, June 1847-June 1848

Mississippi

- 1st Regiment of Mississippi Rifles, on 10 companies, June 1846-June 1847
- 2nd Regiment of Mississippi Volunteers, on 10 companies, January 1847-July 1848
- 1st Battalion of Mississippi Rifle Volunteers, on 5 companies, October 1847-June 1848

MISSISSIPPI VOLUNTEERS

Officer Soldiers

Missouri

- Battalion of Missouri Volunteers, on 5 companies (two of infantry and three of light artillery), June 1846-July 1847
- Battalion of Missouri Mounted Volunteers, on 4 companies, August 1846-August 1847
- 1st Regiment of Missouri Mounted Volunteers, on 8 companies, June 1846-June 1847
- 2nd Regiment of Missouri Mounted Volunteers, on 5 companies, May 1847-October 1848
- 3rd Regiment of Missouri Mounted Volunteers, on 10 companies, July 1847-October 1848
- Battalion of Missouri Mounted Volunteers (new), on 5 companies, August 1847-November 1848
- Battalion of Missouri Mounted Volunteers (new), on 5 companies, September 1847-October 1848
- Battalion of Santa Fe Mounted Volunteers (New Mexico territory), on 4 companies, July 1847-October 1848

New Jersey

- Battalion of New Jersey Volunteers, on 4 companies, September 1847-July 1848

New York

- 1st Regiment of New York Volunteers (aka California Guard/California Regiment/New York Legion), on 5 companies, August-October 1848
- 2nd Regiment of New York Volunteers, on 10 companies, November 1846-August 1848

North Carolina

- Regiment of North Carolina Volunteers, on 10 companies, January 1847-August 1848

Ohio

- 1st Regiment of Ohio Volunteers, on 10 companies, June 1846-June 1847
- 2nd Regiment of Ohio Volunteers, on 10 companies, June 1846-June 1847
- 3rd Regiment of Ohio Volunteers, on 10 companies, June 1846-June 1847
- 4th Regiment of Ohio Volunteers, on 10 companies, May 1847-July 1848
- 5th Regiment of Ohio Volunteers, on 10 companies, September 1847-July 1848
- 1st Independent Company of Mounted Men, June 1847-August 1848
- 2nd Independent Company of Foot, October 1847-July 1848
- 3rd Independent Company of Foot, October 1847-July 1848

Pennsylvania

- 1st Regiment of Pennsylvania Volunteers, on 10 companies, December 1846-August 1848
- 2nd Regiment of Pennsylvania Volunteers, on 12 companies, January 1847-July 1848

South Carolina

- Regiment of South Carolina Volunteers, on 10 companies, December 1846-July 1848

Tennessee

- Regiment of Tennessee Mounted Volunteers, on 10 companies, June 1846-May 1847
- 1st Regiment of Tennessee Volunteers, on 12 companies, May 1846-May 1847
- 2nd Regiment of Tennessee Volunteers, on 8 companies, June 1846-May 1847
- 3rd Regiment of Tennessee Volunteers, on 10 companies, October 1847-July 1848
- 4th Regiment of Tennessee Volunteers, on 9 companies, November 1847-August 1848
- 5th Regiment of Tennessee Volunteers, on 11 companies, December 1847-July 1848

Texas

- 1st Regiment of Texas Mounted Rifle Volunteers, on 10 companies, June-October 1846
- 2nd Regiment of Texas Mounted Rifle Volunteers, on 10 companies, June-October 1846
- Regiment of Texas Rifle Volunteers, on 10 companies, May-August 1846
- Regiment of Texas Mounted Volunteers, on 7 companies, July-September 1846
- Battalion of Texas Mounted Volunteers, on 4 companies, March 1847-June 1848
- Regiment of Texas Mounted Volunteers (new), on 10 companies, April 1847-May 1848
- Regiment of Texas Mounted Volunteers (for frontier defence), on 10 companies, May-June 1847
- 1st Independent Company of Mounted Volunteers, September 1845-June 1846
- 2nd Independent Company of Rangers, September 1845-July 1846
- 3rd Independent Company of Mounted Volunteers, October 1845-September 1846
- 4th Independent Company of Mounted Volunteers, May-September 1846
- 5th Independent Company of Mounted Volunteers, July-October 1846
- 6th Independent Company of Mounted Volunteers, July 1846-July 1847
- 7th Independent Company of Foot Volunteers, August 1846-January 1847
- 8th Independent Company of Rifles, September 1846-July 1847
- 9th Independent Company of Volunteers, October 1846-October 1847
- 10th Independent Company of Volunteers, October 1846-October 1847
- 11th Independent Company of Spies, January-July 1847

Virginia

- Regiment of Virginia Volunteers, on 14 companies, December 1846-August 1848

Some of the volunteer units listed above had a very peculiar history, especially those coming from the "youngest" American states. Arkansas was admitted to the Union in June 1836 and very soon its militia was "activated" for military service. In that particular moment the regular forces of the US Army were heavily involved into the Second Seminole War that was ravaging Florida; in addition, in Texas the revolutionaries of Sam Houston were fighting against the expeditionary force of Santa Anna to defend the independence of the newly-born Republic of Texas.

Since the federal government removed most of its regulars from the border with the native territories in order to deploy them on other fronts, the Governor of Arkansas was required to recruit a force of 1,000 volunteers coming from the ranks of his state militia. These were to be organized into a regiment with ten companies, but the recruiting operations were not easy and thus it was possible to raise only a battalion with five companies. During the following weeks, however, more companies of volunteers coming from every corner of Arkansas joined the battalion and thus it was possible to transform the latter into a regiment with nine companies. The unit was known as "1st Regiment of Arkansas Mounted Gunmen" and was disbanded in February 1837 when the military menace represented by the Texas Revolution was over. During the emergency, the Governor of Arkansas had started raising also a "2nd Regiment of Arkansas Mounted Gunmen"; this, in reality, always remained a battalion with three companies (increased to five some time later). It was disbanded in April 1837, shortly after the other volunteer unit.

While these events took place, the militia of Arkansas was completely reorganized; among the various states of the Union, the latter was probably the most exposed to foreign menaces and one in which the native presence was extremely significant. As a result, the Arkansas militia was well organized and quite numerous since the beginning. According to the structure adopted in 1836, this was to comprise two divisions with three brigades each. There was also a single company/battery of artillery, known as "1st Artillery Company of Arkansas Militia". In 1843 a new "Militia Act" reorganized the militia units of Arkansas, by adding another two brigades to the existing ones (the population of the young state had expanded a lot and thus the total number of militia regiments was increased to 46). As in all the other American states, however, most of these units existed only on paper. When the War with Mexico broke out, the federal government required the formation of a cavalry regiment and of an infantry battalion to the Governor of Arkansas; these, like in 1836, were to substitute the regulars on the border with the native territory. In total, 22 companies of cavalry and 7 of infantry answered to the call of their governor; these were too numerous, however, since only 15 companies were needed. The "Regiment of Arkansas Mounted Volunteers", on 10 cavalry companies, was deployed to Mexico; the "Battalion of Arkansas Volunteer Infantry" with 5 infantry companies, instead, was sent to serve on the western frontier in order to prevent eventual native incursions.

After the bloody Battle of Buena Vista, most of the volunteers from Arkansas returned home after completing their 12 months of enlistment. Like in many other American states, those few volunteers who decided to continue serving with the regular army until the end of the war were assembled into new "independent" companies. All the volunteer horse units deployed by the various states, including those from Arkansas, were not proper "cavalry" corps but "mounted infantry" ones. Their members, in fact, used horses for travelling long distances but usually dismounted to fight. They were armed with infantry rifles and did not have cavalry sabers.

The California Guard, the first US militia unit formed in California in 1849

The volunteer unit raised in California, commonly known as "California Battalion", had a very peculiar history. When the war between United States and Mexico broke out, the local community of American colonists (which was quite large like the Texan one) revolted against Mexico and proclaimed the independence of California during the so-called "Bear Flag Revolt". The new "Bear Flag Republic", however, existed only for 25 days since it was soon annexed to the United States. US Army Captain John C. Frémont, who was already doing secret cartography work in California, was ordered to take over the command of the Californian Revolt.

As the only US military officer in the region, Frémont commanded a small force of 30 cartographers and 30 irregular scouts/hunters; since these were too few to secure the independence of California, he had no choice but to recruit in a very short time a large number of Californian volunteers. These were formally organized into the "California Battalion" on 23 July 1846; the new unit comprised several natives who had joined the "Bear Flag" cause as well as all Frémont's original men (many of whom acted as officers in the new battalion). The main task of the Californian corps was that of helping the landing parties of the US Navy's "Pacific Squadron" to conquer the major coastal cities and ports of California. The region did not have significant Mexican garrisons and thus it was quite simple for the Marines coming from the US warships and for Frémont's men to occupy most of California without suffering heavy losses. In this phase of the operations, the US Pacific Squadron comprised three frigates with a crew of 480 men each plus four sloops with a crew of 200 men each. Of the 2,240 sailors and Marines who operated along the coasts of California, more or less 400 were employed as "naval infantry" to occupy and garrison the major ports of the region; these, of course, were progressively substituted by the men of the "California Battalion". The latter, in fact, was gradually expanded thanks to the arrival of numerous new American settlers who moved to California soon after it proclaimed its independent from Mexico. With the new recruits Frémont was able to expand the numerical consistency of his unit, which now comprised eight companies of infantry and one of artillery.

The infantry companies were lettered from A to H, while the artillery one did comprise only two small guns. The friendly natives who supported Frémont were all part of Company H. In addition to the above, the unit was supplemented by a large number of irregulars: these were outriders, hunters, couriers and scouts including the famous Kit Carson. At the outbreak of the war with Mexico, in order to launch a "secondary" campaign against the enemy territories of New Mexico and California, the US government had assembled several military units at Fort Leavenworth in Kansas. These were under command of Colonel Stephen W. Kearny and made up the so-called "Army of the West". The latter comprised the following regular and volunteer corps: five companies of the 1st Dragoons (300 men), two infantry companies from the Battalion of Missouri Volunteers (100 men), three artillery companies from the Battalion of Missouri Volunteers (150 men, with twelve 6-pdr guns and four 12-pdr mountain howitzers), the 1st Regiment of Missouri Mounted Volunteers (with 850 men in eight companies), one company of "LaClede Rangers" (100 mounted volunteers from Saint Louis), 50 friendly natives employed as scouts (from the Shawnee and Delaware tribes). The regulars of the "Army of the West" were very few, since most of Kearny's force consisted of volunteers recruited in Missouri. The military units listed above invaded New Mexico and conquered it after a few minor clashes; some time later, they were supplemented by two new units that were sent as reinforcements to Colonel Kearny: the 2nd Regiment of Missouri Mounted Volunteers (with 500 men in five companies) and the Mormon Battalion (with 500 men in five companies). The 3rd Regiment of Missouri

Mounted Volunteers and the Battalion of Santa Fe Mounted Volunteers were sent as additional reinforcements at a later date. Kearny's "Army of the West", after securing its possession of New Mexico, moved to California where it joined forces with Frémont's California Battalion and completed the "liberation" of California. During the following months, in March 1847, new reinforcements arrived in California from the sea: the 1st Regiment of New York Volunteers (500 men in five companies) and one battery from the regular 3rd Artillery Regiment. These units travelled by sea from the Atlantic coast to the Pacific one, but arrived when California was already in American hands.

Monument dedicated to the memory of the Mormon Battalion

The Mormon Battalion that took part to the war against Mexico is the only religion-based unit ever existed in the military history of the United States and thus deserves some special attention. This corps, which had senior officers coming from the US Army and junior ones elected from the enlisted men, was able to make an epic march of nearly 2,000 miles from Council Bluffs in Iowa to San Diego in California and opened a new southern wagon route to California; as a result, it played a very important role in the early colonization of the American west. Before 1845, being under continued religious persecution, the members of "The Church of Jesus Christ of Latter-day Saints" were searching for a new land in the west where they could live in peace. In particular, they wanted to move to the Salt Lake Valley located west of the Rocky Mountains; to do this; however, they needed permission and some support from the federal government. The Mormons had just crossed the Mississippi River in order to escape from Illinois and were now temporarily settled in Iowa. The latter was not yet a state but a simple "territory", where the Mormons could remain for some time. While these events took place, the hostilities between USA and Mexico broke out. When the Mormons sent some of their representatives to Washington in order to ask for the help of the central government, President Polk had the idea of employing this religious community as part of the military forces that would have occupied the Mexican territories of the south-west. In exchange for forming a military unit with some hundreds of volunteers, the Mormons received permission to move from Iowa towards the Rocky Mountains in order to create their new settlement. This way President Polk hoped to resolve the religious problems related to the persecutions of the Mormons and tried to transform them into an "organic" component of the American society. Initially the leaders of the religious minority were contrary to the creation of a military unit entirely formed by Mormons, but later they understood that this could have been a great opportunity for them to have a public recognition from the central government (which had always ignored their requests). All the soldiers of the battalion were given a uniform allowance of $42 each; since these were not used to purchase uniforms, the Mormons were able to put aside a large sum of money that was employed to buy wagons and all the other materials needed to build a new settlement in the west. After three weeks the recruiting operations ended and the five companies needed to form a battalion were assembled. The Mormons marched for weeks over some of the most arduous terrain of the American continent, building a new wagon road during their desperate march. On 29 January 1847, without fighting a single battle against the Mexicans, the battalion arrived in San Diego and completed its trans-continental journey.

One of the best volunteer units that took part to the US-Mexican War was the 1st Regiment of Mississippi Rifles which commander, Jefferson Davis, would later become the President of the CSA during the American Civil War. Mississippi was one of the states where enthusiasm for the outbreak of the hostilities was extremely high and where large numbers of volunteers had enlisted; in a few days, in fact, 25 companies were raised in the state instead of the 10 needed ones. Davis equipped his men with excellent percussion muskets and made them one of the best trained volunteer corps of the US Army. Jefferson Davis' regiment fought with great valour at the important Battle of Buena Vista (where it was decisive) and at the later Battle of Monterrey. Thanks to the great popularity gained during the US-Mexican War, Jefferson Davis made a brilliant political career after the end of the hostilities. Another important unit of volunteers that took part to the campaign in Mexico was the Regiment of South Carolina Volunteers, commonly known as "Palmetto Regiment". The latter was the first American military unit to raise its own colours over Mexico City and took part to several of the most important military engagements of the US-Mexican War. The unit received some excellent training from the South Carolina Military Academy, an important military institution created by the State of South Carolina in 1842 and still active today (it is commonly known as "The Citadel").

Finally, the volunteer military units raised in Texas deserve some space because of their very peculiar history. During the years 1836-1846, this state was a fully independent country known as "Republic of Texas" or "Lone Star Republic" because of its flag with a single star. After the victorious Battle of San Jacinto, in fact, Texas seceded from Mexico but was not admitted into the Union. As a result, since 1835, the Texans had their own independent army and navy with specific uniforms and weapons; this is something unique in American history, since no other states except for California and Hawaii have ever deployed an independent military force before being annexed by the federal government. In November 1835 the first regular/permanent Texan Army was organized with the following units: one regiment of line infantry with two battalions and one regiment of artillery with two battalions. The two regiments, together, formed a foot brigade. Each of the four battalions consisted of five companies and each company did comprise 56 men. Infantry companies included one captain, one first-lieutenant, one second-lieutenant, four sergeants and four corporals; artillery ones had the same internal establishment, but with an additional third-lieutenant.

Jefferson Davis' Mississippi Volunteers at Buena Vista

In December 1835 the Texan Army was expanded with the addition of a "Legion of Cavalry", consisting of two squadrons with three companies each. Single companies had 60 troopers each, but had the same internal structure of the foot ones. Soon after independence, in 1836, Sam Houston reorganized the Texan Army by raising another three infantry regiments; the four line regiments were assembled into two brigades with two regiments each, while the single artillery regiment became an independent unit. In addition, brigade and regimental staffs were organized for the first time. Very soon, however, the scarce discipline of the Texan regulars and the economic problems of the Lone Star Republic led to an abrupt reduction in the overall establishment of the Texan Army. By the end of 1837, according to Sam Houston's will, the whole permanent force had been reduced to just a few hundred soldiers who were garrisoned at San Antonio and Galveston. On 10 December 1839 Mirabeau B. Lamar became the new President of Texas and this soon had some important consequences for the army. Lamar, in fact, wished to transform Texas into an expanding country and had plans to conquer most of Mexico's northern territories. As a result, to put in practice his political ideas, the new President of Texas needed a large regular army. A new "extra-large" regiment of line infantry was organized, having 15 companies and being known as "Frontier Regiment" or "First Regiment of Infantry". This unit, deploying a total of 840 men, was stationed in several newly-built forts that were constructed along the so-called "Military Road"; the latter ran from the Red River to the Nueces River and connected several fortified settlements. Despite its official denomination, the "First Regiment of Infantry" did comprise also some cavalry companies. Lamar's ambitious plans, however, never became reality and the "Frontier Regiment" always continued to suffer from desertions and mutinies. Of the 15 companies originally planned, only ten were effectively organized (with a total of 560 men). In December 1841 Lamar ended his mandate and was replaced by Sam Houston who, during 1842, completely disbanded the remaining companies of the "Frontier Regiment".

As a result, when the Republic of Texas was annexed by the United States on 19 February 1846, the Texan regular military forces were no longer in existence since several years. The regular military forces of Texas did comprise also a small Marine Corps, which was formed on 14 January 1836. During the Texas Revolution this included just a few naval infantrymen, who were not organized into larger units but who were simply assembled into small detachments that were embarked on the four warships of the Texan Navy. By the end of 1837, due to the disappearance of the military menace represented by the Mexicans, the Texan Navy was formally abolished together with its small Marine Corps. During 1839 a "new" Texan Navy was formed, also this time comprising some small detachments of Marines. The latters, as usual, were mostly employed as sharpshooters and boarders on the few warships of the Texan Navy. By late 1843 all the sailors and marines of Texas were placed on the inactive list; as a result, by the time of the US annexation, the Texan Marine Corps existed only on paper. In addition to the regular units described above, Texas also deployed a large militia and several para-military "police" companies of Rangers. According to the legislation passed on 6 December 1836, every able-bodied man aged between 17 and 50 years was to be enrolled in the militia; as a result, it was possible to organize two divisions of militiamen with two brigades each. A single brigade had two infantry regiments with two battalions each; a single battalion was to have five companies. According to the 1846 militia legislation, battalions and companies were to muster twice a year for training, while regiments were to be assembled only once a year for three days of intensive drill. On paper this militia structure was absolutely formidable, but the Texans never had the needed economic resources to make it work. Differently from Lamar, Houston believed that the militia could have been much more disciplined and efficient than a permanent corps of regulars; like those of Lamar, however, also his plans mostly remained theoretical. A new Militia Act was passed in 1837, with the objective of reorganizing the militia units; according to this, the militiamen of Texas would have been assembled into a single division with four brigades. Also this new organization remained only on paper, although it was formally retained until Texas was annexed by the Union.

The famous Texas Rangers can trace their origins back to 1823, when Stephen F. Austin organized some small and irregular "armed groups" in order to protect the lives and properties of the 70 American families that had settled in Texas after Mexico became an independent nation. The frontier settlements of Texas were greatly exposed to the attacks of the natives and bandits; since the new Mexican government had no intention to defend them, Austin and other colonists had to organize their own para-military forces that could "range over the countryside" (hence the unit's name). From a formal point of view, the Texas Rangers were constituted on 24 November 1835 soon after the outbreak of the Texas Revolution. The new corps was to consist of three independent companies with 60 men each; very soon, however, it was expanded up to more than 300 men. During the military operations against Mexico, the Rangers performed a series of different military functions: they acted as scouts, spies, couriers and guides. Most of them were battle-hardened frontiersmen who did know their territory extremely well and who were used to very harsh conditions of life. Being equipped as mounted infantrymen, the Rangers could move very rapidly also on broken terrain and were extremely efficient in contrasting native incursions. After Texas achieved independence, the Rangers were reduced to a single company; when Sam Houston was re-elected President of Texas in late 1841, however, it was decided to increase the total number of Rangers to 150. The latters, in a time during which the regular army was being disbanded and the militia was still mostly inefficient, were the only force that could patrol the immense borders of Texas.

Mounted Volunteer from Texas

In addition, maintaining the Rangers on active service had very little costs for the state. Commanded by Captain John Coffee Hays and armed with excellent Colt revolvers, the Rangers were the only "proper" military unit existing in Texas by the time of the US annexation in 1846. During the US-Mexican War several independent companies of Rangers, formally known as "mounted volunteers", were mustered into federal service and soon proved their effectiveness as counter-guerrilla fighters and guides for the US Army. Thanks to their great knowledge of the territory over which the campaign was fought and to their hit-and-run tactics, the Rangers were the only American soldiers who could contrast the Mexican irregular "guerrilleros" by using their same tactics and who could act as explorers or spies for the advancing regular columns.

The Mexicans developed a true "reverence" for the Rangers, who were commonly known as "Los Diablos Tejanos" (i.e. "The Texan Devils") south of the Texan border. The Rangers were the "eyes and ears" of the US Army: superbly mounted and armed to the teeth with a large variety of weapons, they proved to be ruthless and lethal on several occasions. The pre-war Rangers who were commanded by Hays were not disbanded when Texas was annexed by the USA; instead, they formed the core of the new 1st Regiment of Texas Mounted Rifle Volunteers that was raised soon after the outbreak of the hostilities with Mexico. This, together with the 2nd Regiment of Texas Mounted Rifle Volunteers, was one of the best volunteer units deployed during the US-Mexican War. Both regiments comprised substantial numbers of former regular soldiers of the Texan Army and thus made up a sort of "semi-independent" Texan Army attached to the US one. The two cavalry regiments, in fact, were assembled into an independent division that was under command of James Pinckney Henderson (first Governor of Texas). The Texan volunteer cavalrymen were to furnish their own weapons and uniforms and were raised for six months of service.

The soldiers of the 1st Regiment, which was commanded by the same Hays, came from the western part of Texas; those of the 2nd Regiment, instead, came from the eastern part of Texas. The staff of each regiment did comprise one colonel (commanding the unit), one lieutenant-colonel, one major, one adjutant, one sergeant-major, one quartermaster-sergeant and two principal musicians. Each regiment included ten companies, which had the following internal composition: one captain, one first-lieutenant, two second-lieutenants, four sergeants, four corporals, two musicians and 80 troopers. The Texan mounted volunteers were mostly employed to patrol their army's line of communications as well as to gather intelligence about enemy forces. The new independent companies of Texan volunteers formed during the conflict continued to perform the same duties of the original two regiments; they even included a special company of "spies". Under many points of view, Texas was the single state that gave the most significant contribution to the American cause.

US Model 1839 Forage Cap, the standard headgear worn by most of the volunteer units.

US troops in a Mexican city

US infantry officers in campaign dress (left) and artilleryman in parade dress (right)

General Zachary Taylor (left) and General Winfield Scott (right)

Colonel Stephen Watts Kearny (left) and Captain John Charles Frémont (right)

Captain Robert Edward Lee (left) and Lieutenant Ulysses Simpson Grant (right)

Second-Lieutenant Thomas J. Jackson (left) and Second Lieutenant George B. McClellan (right)

Weapons

In 1845 the US Army was equipped with excellent weapons, which performances were comparable to those of the arms employed by the major European armies of the time. The artillery, in particular, could deploy modern field pieces that proved to be extremely important for the final American victory. Like most of the European armies, also the US one had adopted percussion-lock weapons by the outbreak of the hostilities with Mexico. Until 1842, the standard infantry weapon of the US Army was the M1835 flintlock musket; this was an exact copy of the French M1822 infantry musket and proved to be a good weapon. It was a smoothbore musket, with a 0.69 caliber and a length of 57.75 inches. The overall weight of the M1835 musket was about ten pounds, while length of the barrel was 42 inches. In 1842 the US Army adopted a new musket, which was identical in size and appearance to the M1835 one but which had a percussion-lock system instead of a flintlock one. The percussion cap made the M1842 musket much more reliable and easier to reload than its immediate predecessor. The standard average of fire for the M1835 musket was of two rounds per minute, while that of the new M1842 was of three rounds per minute. The level of accuracy of the two muskets, instead, was practically the same since they were both smoothbores. Both muskets had only 10% accuracy at a range of 100 yards under ideal conditions. As a result, infantry tactics of the US Army prescribed to deliver massed volleys of fire at short range into the enemy before launching a frontal bayonet assault. When the war with Mexico broke out, only part of the American infantry companies had received the new percussion weapon and thus many of them went to the theatre of operations with flintlocks that were comparable to the ones used by the Mexicans. By 1845 most of the new percussion muskets that were needed to re-equip the US infantry had already been produced, but there had been no enough time to train most of the companies in the use of the new musket.

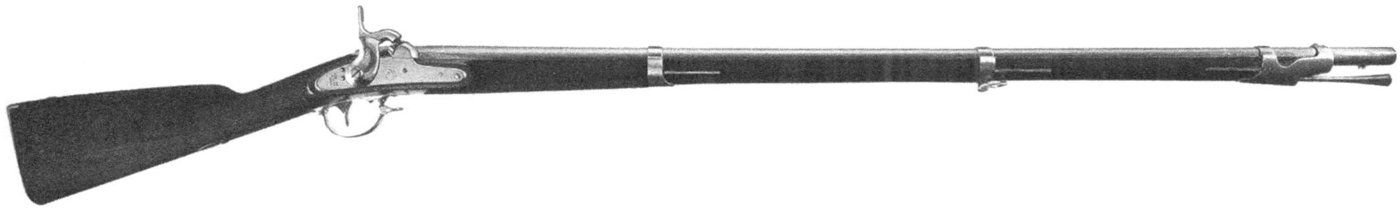

M1842 Springfield Musket

In 1841, for its light infantry units, the US Army adopted a new percussion rifle commonly known as "Mississippi Rifle" because it was soon adopted by the famous Mississippi Volunteers commanded by Jefferson Davis. This was a real precision weapon, probably the best employed until that date by the US Army. The M1841 musket had a complessive weight of nine and one half pounds and a length of 48 inches; the rifled barrel was 33 inches long and had a caliber of 0.54. The new rifle took longer time to be reloaded if compared with the M1842 smoothbore musket, but had an adjustable rear sight and was quite accurate up to 500 yards. The entire "Regiment of Voltigeurs and Foot Riflemen" was equipped with this weapon. Bayonets for the

M1835, M1842 and M1841 muskets all had the same characteristics: they had a length of about 18 inches and a triangular-shaped blade. Bayonets were fixed to the underlug at the end of the barrel. Sometimes the infantrymen could also use sabre-bayonets with brass hilts (measuring more or less 22 inches long).

M1841 Mississippi Rifle and Walker Colt Revolver

The main weapon of the US cavalry was the M1840 cavalry saber; this had a blade of polished steel that was about 1.25 inches wide at the hilt and about 36 inches long. The overall length of the saber was 38.5 inches. The enlisted men's version of the saber had a brass hilt with wooden grips covered in leather and wrapped with brass wire. The officers' version of the saber had some additional decorative elements, like a hilt of gilded brass and the letters "US" reproduced on both sides of the blade. This kind of cavalry weapon, which was a quite exact copy of the French M1822 light cavalry saber, was employed also by the "Regiment of Mounted Riflemen". The two regiments of US Dragoons employed two different models of single-shot pistols during the war with Mexico: the M1836 "Johnson" flintlock pistol (with a caliber of 0.54 and a smoothbore barrel 8.5 inches long) and the M1842 percussion pistol (with same caliber and smoothbore barrel of the previous model). Like the M1842 musket, when the war broke out only limited numbers of the new percussion pistol had been received by the US troops. The "Regiment of Mounted Riflemen" was initially equipped with M1836 flintlock pistols, but these were replaced during 1847 by new "Walker" Colt revolvers. These were extremely innovative weapons of "new generation", which soon became the favourite short firearms of the deadly Texas Rangers. The "Walker" Colt was a five-shot revolver of 0.44 caliber, with a complessive weight of 4.5 pounds and a barrel that was 9 inches long. The revolver invented by Samuel Colt was incredibly efficient and easy to use; in addition, it could fire five times before being reloaded and this made it clearly superior to any existing pistol. The US Dragoons also carried a short cavalry carbine that had been created specifically for them: this was the M1843 "Hall" breech-loader, a 0.54 caliber percussion weapon that was developed from the M1841 light infantry musket. Differently from the latter, however, the M1843 cavalry carbine had a smoothbore barrel with a length of 21 inches. The chamber and lock of the "Hall" carbine could be easily removed by unscrewing a single screw; this allowed the dragoons to carry it as a concealed handgun when on the march. The elite "Regiment of Mounted Riflemen" used the M1841 infantry rifle and not the M1843 cavalry carbine.

During the decade 1835-1845, all the iron field pieces of the US Artillery were gradually substituted by new bronze ones. At the same time, the old gun carriages dating back to the Napoleonic period were replaced by new ones copied from the contemporary British box trails. The

US M1840 field gun of 6 lb.

US Artillery deployed three different types of pieces: guns (used for high-velocity fire with flat trajectory), howitzers (used for heavy-weight fire with arcing trajectory) and mortars (used for heavy-weight fire with steep trajectory). The guns had long barrels, the howitzers had short barrels and the mortars were mostly employed during siege operations. The US field guns included M1840 guns of 6 lb. and 12 lb. as well as M1840 howitzers of 12 lb. and 24 lb.; all these had been designed after the return of the American military mission that had been sent to Europe in order to study artillery systems. In addition to the above, there was also a 12 lb. M1835 mountain howitzer, which was specifically designed for use on broken terrain. Siege pieces included M1840 12 lb., 18 lb. and 24 lb. guns as well as 8-inch siege howitzers and four different models of mortar (8-inch and 10-inch light ones, 16-inch stone ones and 24 lb. heavy ones). Fortress pieces included the following models: M1840 32 lb. and 42 lb. guns, M1819 24 lb. guns, M1829 32 lb. guns, M1831 42 lb. guns, 8-inch and 10-inch howitzers, 10-inch and 13-inch heavy mortars. Of all the artillery pieces listed above, only some were effectively employed during the war with Mexico. The 6 lb. field gun had a smooth-bore diameter of 3.67 inches and a weight of 880 pounds; the 12 lb. field gun had a smooth-bore diameter of 4.62 inches and a weight of 1,800 pounds. The 12 lb. field howitzer had a smooth-bore diameter of 4.62 inches and a weight of 785 pounds; the 24 lb. field howitzer had a smooth-bore diameter of 5.82 inches and a weight of 1,320 pounds. Finally, the 12 lb. M1835 mountain howitzer was a light piece with a smooth-bore diameter of 4.62 inches and a weight of only 210 pounds. The entire M1835 mountain howitzer could be rapidly disassembled and easily packed on three horses for transport. The American artillery did employ also an "updated" version of the old British Congreve rockets, having a 9 lb. projectile with a diameter of 2.5 inches. The US artillery also had a certain number of Hales rockets, developed by the American inventor William Hale; these had a 6 lb. projectile with a diameter of 2.25 inches. A heavier version of the Hales rockets did exist, with a 16 lb. projectile having a diameter of 3.25 inches. All the Congreve and Hales rockets were fired from portable stands or simply propped up on the ground. The "Company of Sappers, Miners and Pontoniers" was equipped with a specific short weapon, known as M1847 "Sappers and Miners Musketoon". This was a 0.69 caliber smoothbore short musket having a complessive length of 41 inches. The musketoon had a brass-handled sword-bayonet that was fixed on the right side of the barrel by a bayonet stud.

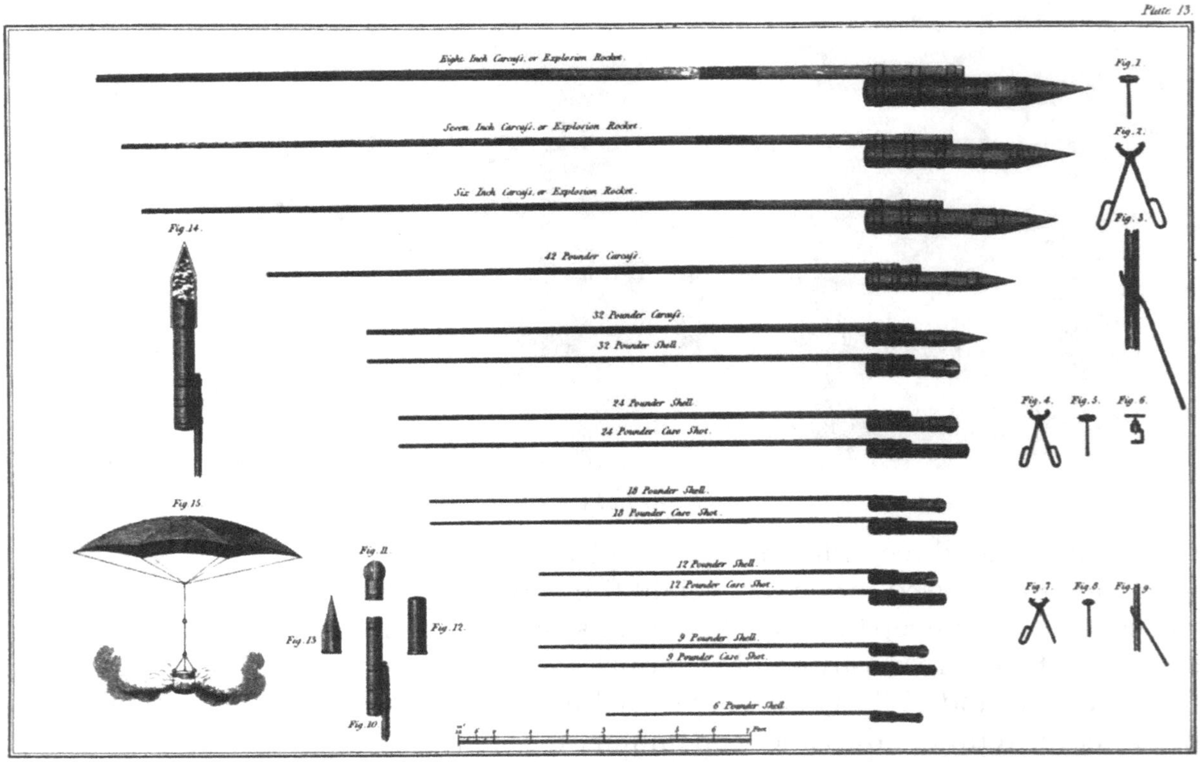

Various models of Congreve Rockets

The Mexican Army of 1821-1848

On 27 September 1821 the former "Viceroyalty of New Spain" became independent from Spain, with the new denomination of "Mexican Empire". The latter was guided by Augustin de Iturbide, who had been the leader of the independentist movement during the final phase of Mexico's Independence War (1810-1821). The Mexican struggle for freedom had been extremely long and complicated: two important revolutionary movements, one guided by Hidalgo and the other by Morelos, had been suppressed by the Spanish military forces respectively in 1811 and 1815. The insurgents fought with great courage and determination, but since they had no military organization to speak of they were no match for the Spanish regular military forces. These comprised two categories of troops: the "metropolitan" units coming from Spain that were sent from the motherland in order to crush the rebellion and the "colonial" units recruited from the Mexican population. The latter were well trained and equipped, having more or less the same standards of service of the "metropolitan" Spanish units. The Spanish colonial units were known as "fijas" or "fixed", since they were formed by professional soldiers and were usually stationed in a specific location (many of them had a distinct "garrison" character). In addition to the metropolitan and fixed units, there was also the local militia but this was of little military use since it lacked effective training and proper equipment.

Until the early months of 1821 the social elites of Mexico, who had the leadership of the "fijas" military units, remained loyal to the Spanish Crown and continued to fight with all their resources against the members of the lower social classes (who made up the majority of the insurgents across the whole Mexican territory). At a certain point, however, the elites of the country understood that they could benefit a lot from the eventual independence of Mexico and started to doubt about Spain's possibility to dominate over the American colonies. The Spanish Crown was becoming increasingly weak, also because the same Spanish territory was ravaged by internal revolutions; most of South America had already become independent from the motherland and thus it would have been quite difficult for Spanish Mexico to have positive relations with the new republics of Latin America in the near future. As a result of the above Augustin de Iturbide, who was one of the main military leaders of the "fijas" military units and who had fought with great distinction against the insurgents for several years, decided to change side and join forces with the lower social classes of the population in order to achieve the independence of Mexico. This decision completely changed the military situation on the field: by uniting the "fijas" military forces and those of the insurgents under his unified command, Iturbide was able to achieve a global superiority over the metropolitan units that were fighting against Mexican independence.

After Morelos' death, the leadership of the "popular" insurgent forces had been assumed by Vicente Guerrero; the latter did not have trust in Iturbide, but finally decided to join his forces with those of the colonial military commander. The new combined military force that was created after this political move did comprise several excellent "fijas" units but also many insurgent corps having a lower degree of organization. It soon became known as "Ejército Trigarante" or "Army of the Three Guarantees", since it was created according to a "Plan of the Three Guarantees" (also known as "Plano de Iguala") that was signed between Iturbide and

Guerrero. The "Three Guarantees" on which the future of Mexico would have been based were religion, independence and unity: religion, because Catholicism was something that united the lower social classes with the upper ones; independence, because Mexico did not want to remain in the Spanish Empire; unity, because the Mexican national territory was extremely large and thus would have needed a federal form of government in order to be ruled (avoiding the possibility of secessions). Thanks to the global superiority of the "Ejército Trigarante", Mexican independence was achieved shortly after the signing of the "Plano de Iguala". The Spanish troops abandoned Mexico without putting up a strong resistance and Iturbide became the real master of the new nation.

Iturbide as Emperor of Mexico

After entering at the head of his army in Mexico City, Augustin de Iturbide was proclaimed "Emperor of Mexico" and thus became the first military dictator in the history of the country. Many independentist leaders, like the same Guerrero, hoped to adopt a republican form of government for their newly-independent country: this, however, was not possible because thanks to the support of the army Iturbide had complete control over Mexico City. By now the insurgent leaders had only scarce irregular forces under their command; in addition, it would have had very little sense for them to start a civil war just a few days after the defeat of the Spanish troops. As a result, Iturbide could rule as the sole master of Mexico until 1823; during this short period, the territory of the Mexican Empire did comprise most of Central America and included the following states: Guatemala, Nicaragua, Honduras, Costa Rica and El Salvador. In 1823, however, the Mexican republicans revolted against Iturbide and a large part of the Mexican Army marched against the Emperor. The rebels were guided by several leaders who had fought in the War of Independence, but most notably by General Santa Anna who had already distinguished himself on several occasions. Iturbide was forced to go in exile and the Mexican Empire became a federal republic. Thanks to the important role played during the 1823 revolution, Santa Anna became one of the most prominent Mexican politicians and acquired the loyalty of the military forces. The events of 1823, however, did not have only positive consequences for the Mexicans: the provinces of Central America listed above, in fact, proclaimed their independence as the "Federal Republic of Central America" (which lasted until 1838, when the five provinces all became independent republics). A new era of Mexican history had begun.

On 27 September 1821, when the "Ejército Trigarante" entered Mexico City, the Mexican military forces comprised a total of 16,134 soldiers: 7,416 infantrymen, 7,955 cavalrymen and 763 artillerymen with 68 guns of different calibre. A good number of these were regular soldiers from the "fijas" units, who were extremely loyal to Iturbide; the remaining ones were all semi-regular insurgents, mostly equipped as light cavalrymen. It is interesting to notice, in fact, that the cavalry of the new army was more numerous than the infantry. Soon after becoming "Emperor of Mexico", Augustin de Iturbide started to reorganize the "Trigarante Army" in order to transform the latter into a homogenous and effective military force. Basically, the new Mexican Army was to comprise three categories of troops: the regular/permanent units of the line, the corps of the "Milicia Activa" (Active Militia) and the corps of the "Milicia Civica" (Civic Militia). The latter changed denomination in 1845, becoming the new "Guardia Nacional" (National Guard). The permanent units, which were made up of professional soldiers, formed the regular army and comprised a large number of men who had served in the former "fijas" military units. The Active Militia, instead, was the direct heir of the previous "provincial corps" formed by the Spanish government across the whole territory of Mexico. It consisted of militiamen who were called to serve on active duty also during peacetime and who had to perform a series of "auxiliary duties"; for example, most of the military garrisons that were scattered on the Mexican territory were made up of active militiamen and not of regular soldiers (who were employed for more difficult tasks). The Civic Militia was the national reserve into which, according to Mexican law, all able-bodied citizens aged between 18 and 40 years could be enlisted in case of national emergency. Generally speaking, the civic militiamen trained only occasionally and had very poor personal equipment; as a result, in case of war, they would have been of very little military use except for small-scale police units against local insurgents or bandits. Ideally, the Civic Militia would have been formed by volunteers who wished to serve their country in case of need; in practice, however, this "reserve army" that existed only on paper was usually formed by compulsory listing of one man on every hundred. On 10 September 1823 the whole Mexican Army was organized into 17 "Commandancias Generales" (General Commandancies) and 7 "Commandancias Particulares" (Particular Commandancies), in order to control the whole territory of the state; in 1839 the Commandancies were assembled into six territorial divisions. The latter were under control of four General Commandancies: the northern one of the Californias, the central one of Mexico City, the south-western one of Chiapas and the south-eastern one of Yucatan. Some of the most important cities or fortresses of the state had their own independent military governor/commandant. By 1825 the Mexican Army comprised a total of more or less 30,000 soldiers; during the following decades, its standard strength was of 25,000.

Guard units

Before its entrance into Mexico City, the "Ejército Trigarante" of Iturbide comprised two guard units: a battalion of foot grenadiers with 250 men and a squadron of horse grenadiers with 300 cavalrymen. These acted, respectively, as the foot bodyguard and mounted bodyguard of the general. After Iturbide assumed power as Emperor of Mexico, on 7 November 1821, the two small units were transformed into the new "Imperial Guard" and were expanded in order to become regiments. The foot unit, by now known as "Granaderos Imperiales", was augmented to 500 men and was structured on two battalions; the cavalry one, instead, was augmented to 700 men and was structured on four squadrons. In October 1823, when Mexico became a republic, the infantry regiment was disbanded; the cavalry one, instead, continued to exist until September 1825 when it was absorbed into the 2nd Line Cavalry Regiment. Despite their denomination as "horse grenadiers", the cavalrymen of the Imperial Guard formed a light unit having hussar-style uniforms.

During the period 1825-1841, the Mexican Army did not comprise any "guard" unit; it was only on 7 December 1841, in fact, that Santa Anna ordered the creation of a new "Guard of the Supreme Powers". According to Mexican constitution, there was no space for a guard corps in a republican state; Santa Anna, however, wanted to have his own guard units as most of the Latin American dictators of the time. His idea was that of creating a special corps that, at

Foot Grenadiers (left) and Horse Grenadiers (right) of Iturbide

least formally, would have defended the "supreme powers of the republican presidency"; this way he would have been able to justify the existence of a new "praetorian" unit like the one he intended to form. The new corps was known as "Battalion of Grenadiers of the Guard of the Supreme Powers" and was to have an establishment of eight companies with a total of twelve hundred soldiers. Despite official prescriptions, the elite battalion never had more than three hundred officers and men. The new guard unit was originally formed as an Active Militia battalion and became a regular corps only at a later date. It had a minimum height requirement of five and one-half feet and thus was formed by the tallest recruits coming from eight different Mexican states.

In July 1845, soon after Santa Anna temporarily lost power, the Grenadiers of the Guard were disbanded by the republican authorities; when the dictator assumed again power, the battalion was rapidly re-formed with an establishment of 500 men. The Grenadiers of the Guard fought with great courage on several occasions during the war with the United States and their unit was disbanded only in December 1847; the surviving veterans of the battalion were absorbed into the Battalion of Sappers. The new "Guard of the Supreme Powers" organized by Santa Anna did comprise also a cavalry unit, the "Squadron of Hussars of the Guard of the Supreme Powers". This was originally raised, on 3 December 1841, as an Active Militia light cavalry squadron having hussar-style uniforms and equipment. Apparently Santa Anna soon developed a special predilection for this unit, which acted as his mounted bodyguard in Mexico City; as a result, he transformed it into a regular corps and added a second squadron on 1 September 1843. On 27 July 1847, during the war with the United States, another two squadrons were added and thus the unit was transformed into a regiment. Like Iturbide, by 1847, Santa Anna had his own "personal guard" with one infantry unit of grenadiers and one cavalry unit of light horsemen. Like the Grenadiers of the Guard, the Hussars remained extremely loyal to the Mexican dictator until the end and fought with great courage on several occasions during the war against the United States. They were disbanded in December 1847, together with the Grenadiers of the Guard.

GUARD OF THE SUPREME POWERS

Foot Grenadiers
Early parade dress — Definitive parade dress — Service dress

Hussars

Infantry

Since November 1821, the line infantry of the Mexican Army was organized with thirteen regiments with two battalions each; a single battalion consisted of eight companies: one of grenadiers, one of "cazadores" (light infantry) and six of fusiliers. Each infantry company had the following establishment: one captain, one lieutenant, two ensigns, four sergeants, ten corporals, three musicians (drummers for the grenadiers and fusiliers, buglers for the "cazadores") and 83 soldiers. In 1823, with the abolition of the Empire and with the secession of the Central American provinces, the general establishment of the Mexican line infantry was reduced to twelve regiments with a single battalion each; as a result, only twelve of the original twenty-six battalions continued to exist. The number of line infantry units had been excessively high before 1823 because the new Mexican Army had continued to retain all the foot units of the former "Ejército Trigarante". Many of these were under-strength and thus the general reorganization of 1823 was absolutely needed.

The new "lighter" regiments maintained their previous denomination but soon started to be known as battalions due to their smaller establishments. The staff of each regiment/battalion consisted of one colonel (commanding the unit), one lieutenant-colonel, two adjutants, one standard-bearer, one chaplain, one surgeon, one armourer, one paymaster, one drum-major, one bugle-major, one drum-corporal, one bugle-corporal and a small section of eight "zapadores" (sappers). The latters were elite soldiers who were chosen among the strongest and tallest recruits; their main task was that of "opening the way" to the advancing regiment by destroying any artificial or natural obstacle encountered during the march.

Due to economic reasons, on 19 November 1833, the number of line infantry regiments was reduced to ten; in addition, each unit received a peculiar name (until that moment the single units had bore only a progressive number). The ten names were taken from some of the most important figures of Mexico's Independence War: Hidalgo, Allende, Morelos, Guerrero, Matamoros, Aldama, Abasolo, Jimenez, Galeana and Landero. It was with this structure that the Mexican infantry took part to the Texan Revolution of 1836. As the events of that war clearly showed, the consistency of the foot units had to be augmented in order to face the new military menace represented by the United States. As a result, on 16 March 1839, the Mexican line infantry was augmented to twelve regiments with two battalions each; the new denominations adopted in 1833 were abandoned and the units received again progressive numbers. The internal consistency of a single battalion did not change, while that of a single company became as follows: one captain, one lieutenant, two sub-lieutenants, one sergeant 1st class, four sergeants 2nd class, nine corporals, four musicians and eighty soldiers. The regimental staff was changed as follows: one colonel (commanding the whole unit and the 1st Battalion), one lieutenant-colonel, one major (commanding the 2nd Battalion), one sub-lieutenant standard-bearer, two chaplains, two surgeons, two armourers, one drum-major, one bugle-corporal and eighteen sappers (two corporals and sixteen soldiers). In addition to the twelve regiments described above, the Mexican regular infantry did comprise also eight independent companies (garrison units that were employed in specific locations) and one battalion of veterans/invalids structured on four companies (which was made up of pensioners who were garrisoned in Mexico City). In 1840, for the first time, some independent light infantry units were formed: these consisted of three "cazadores" regiments, each having a progressive number. These new units were "elite" ones, since their members underwent some specific training and were all equipped with rifled carbines. A 4th Light Infantry Regiment was created in 1846 by converting the 1st Battalion of the 3rd Line Infantry Regiment into a light unit. The latter was replaced by absorbing the Coast Guard Battalion of San Blas into the permanent line infantry. The light infantry regiments had the same internal structure of the line infantry ones, but their companies had different denominations ("carabiniers" instead of "grenadiers", "cazadores" instead of "fusiliers" and "voltigeurs" instead of "cazadores").

MEXICAN LINE INFANTRY

Sub-lieutenant Soldier "Cazador" Drummer Corporal Standard-bearer

First Regiment

On 8 January 1842 the general establishment of the Mexican regular infantry was augmented with the formation of an independent battalion, known as "Fixed Battalion of the Californias". This was a garrison unit specifically created to defend the vast and isolated territory of California (at that time comprising Upper California and Lower California). The northern part of Mexican California was strongly menaced by the expansionist policies of the United States and was an easy target for any privateer or immigrant who was in search of new lands where to live. According to the original intentions of the Mexican government, the "Californias Battalion" was to be an elite corps formed with excellent recruits; in practice, however, the recruiting operations for the new unit proved to be extremely difficult because no Mexican soldier wanted to serve in such an isolated territory like that of California. As a result, the new battalion was mostly made up of men "recruited" from the Mexican prisons; apparently, most of its officers were criminals who had no choice but to remain in prison for the rest of their life or go to California as soldiers. The 500 infantrymen of the "Californias Battalion" saw very little fighting during the war with the United States and their battalion was finally captured by the Americans on 11 November 1848. In 1843 another independent "fixed" unit was formed: the Battalion "Fijo de Mexico", which members were chosen among the most experienced veterans of the line infantry. The "Fijo de Mexico" was employed as the garrison force of Mexico City and thus had a very important role; it protected all the main buildings of the central government and also acted as a sort of "urban guard" for the vast Mexican capital. During the war with the United States, due to the high casualties suffered, the original twelve line infantry regiments were strongly reduced in their numbers; by December 1847 there were enough soldiers to form only twenty "temporary" battalions, which were later consolidated into eight regiments with two battalions each before the end of the conflict (November 1848).

MEXICAN LINE INFANTRY

2nd regiment | 3rd regiment | 4th regiment | 5th regiment | 6th regiment | universal summer dress

The regular or "permanent" infantry was supplemented by the foot units of the Active Militia; these, during the period taken into account, were always known by the name of their city or province in order to underline the "local" status of these reserve corps. By 1826, the infantry units of the Active Militia did comprise a total of twenty battalions plus several independent "garrison" companies. These battalions and companies had the same internal structure of the permanent ones, but most of them had inferior training and equipment. In June 1840, after the experiences of the Texan Revolution, the infantry of the Active Militia was reorganized on the following units: three regiments, numbered from 1 to 3, with two battalions each; eight battalions, numbered from 4 to 11; eight independent "garrison" companies. In December 1847, in order to consolidate the regular infantry on 20 "temporary" battalions, all the surviving infantry corps of the Active Militia were absorbed into the permanent army and thus changed status.

In addition to the usual Active Militia, the Mexican armed forces did comprise also a special "Coast Guard Militia" that was inherited from the Spanish colonial period. Mexico had an extensive coastline, which was constantly exposed to the naval raids of foreign fleets or of privateers coming from nearby countries; as a result, the Coast Guard Militia had the precise function of patrolling and defending the coasts in case of enemy attacks/incursions. In 1823, with the fall of the Empire, this was reorganized on a total of thirteen battalions; each of the latters comprised only six companies, all of fusiliers. The establishment of the single companies was as follows: one captain, one lieutenant, two sub-lieutenants, one sergeant 1st class, four sergeants

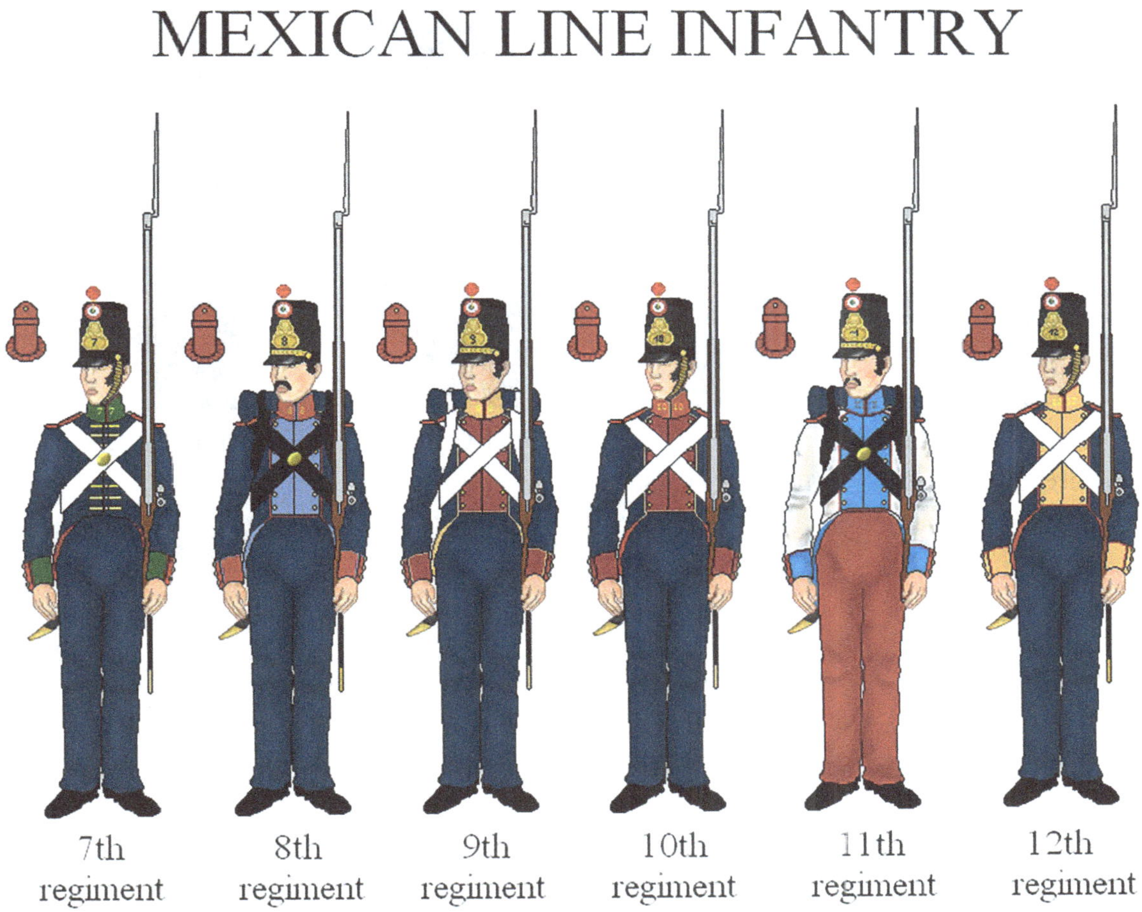

2nd class, six corporals 1st class, six corporals 2nd class, four musicians (two drummers, one fifer, one bugler) and a varying number of soldiers. The staff of each Coast Guard battalion consisted of one commandant, one captain, two lieutenants and one drum-major. In 1839, with the reform of the regular infantry, Coast Guard battalions received the same internal structure of the "permanent" ones. Like the units of the Active Militia, these corps bore the name of the province/city where they were garrisoned and had a distinct "local" nature. In December 1847 they were all dissolved and absorbed into the permanent infantry like the foot units of the Active Militia.

MEXICAN LINE INFANTRY

First regiment — winter dress, summer dress; Second regiment; Third regiment

MEXICAN LINE INFANTRY

Fourth Regiment | Fifth Regiment | Sixth Regiment | Seventh Regiment

MEXICAN LINE INFANTRY

Seventh Regiment (unofficial dress) | Eighth Regiment | Ninth Regiment | Tenth Regiment

MEXICAN LINE INFANTRY

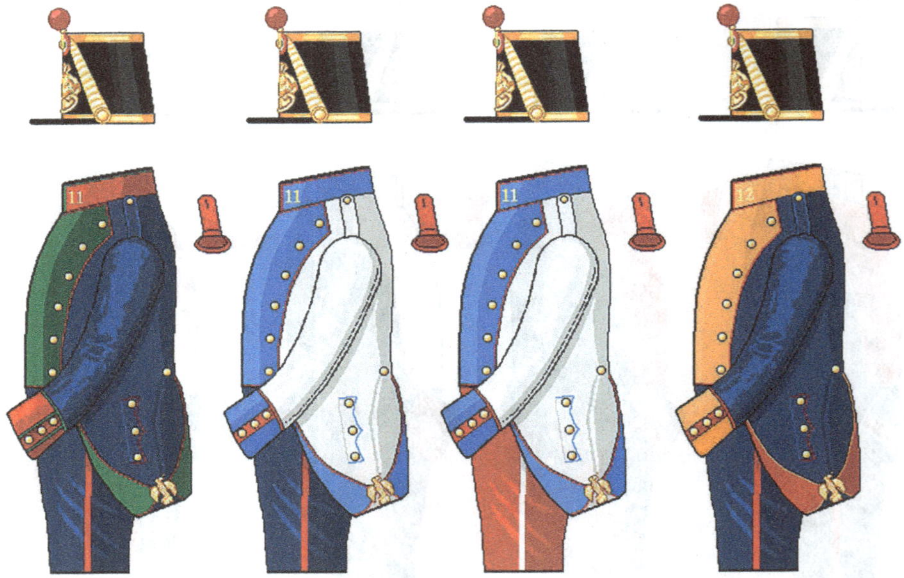

Eleventh Regiment

winter dress summer dress

Twelfth Regiment

MEXICAN LINE INFANTRY

Summer uniforms

New unofficial dress with tunic ("levita")

MEXICAN LINE INFANTRY

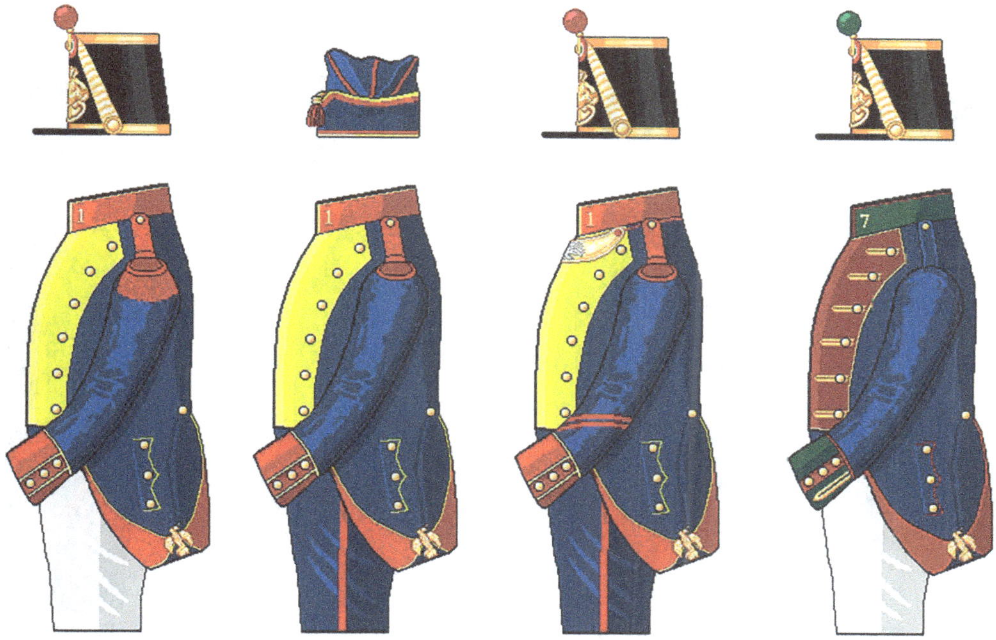

Trumpeter Drummer
First regiment

Standard-bearer

"Cazador"
Seventh regiment

MEXICAN LINE INFANTRY

Captain
Ninth
regiment

Ensign
Second
regiment

Captain with round hat
Fourth
regiment

Captain with kepi
Fifth
regiment

MEXICAN FOOT TROOPS

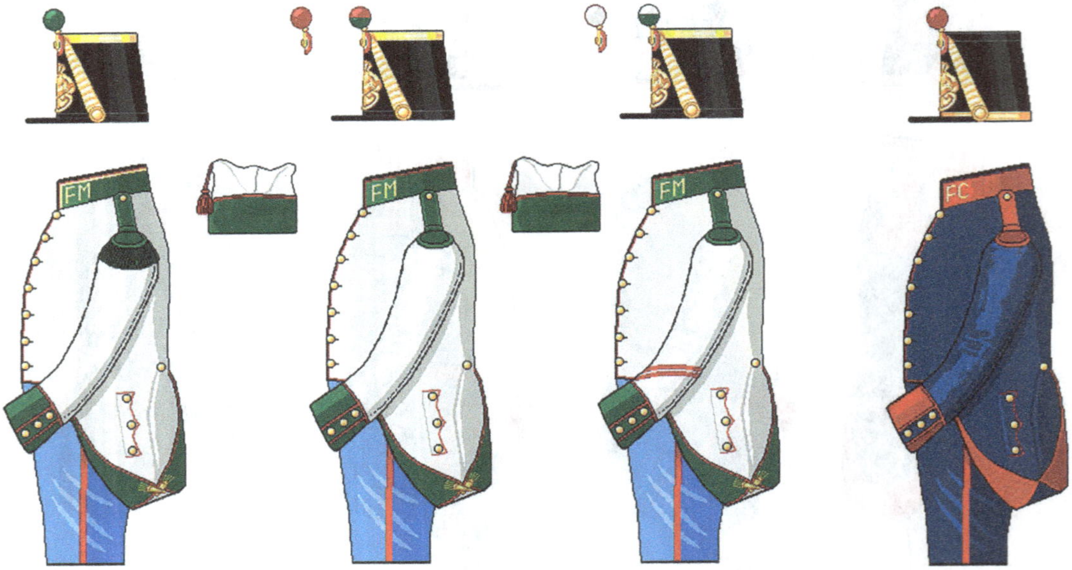

Fusilier — Grenadier — "Cazador"
Fixed Regiment of Mexico

Private
Fixed Regiment of the Californias

MEXICAN LIGHT INFANTRY

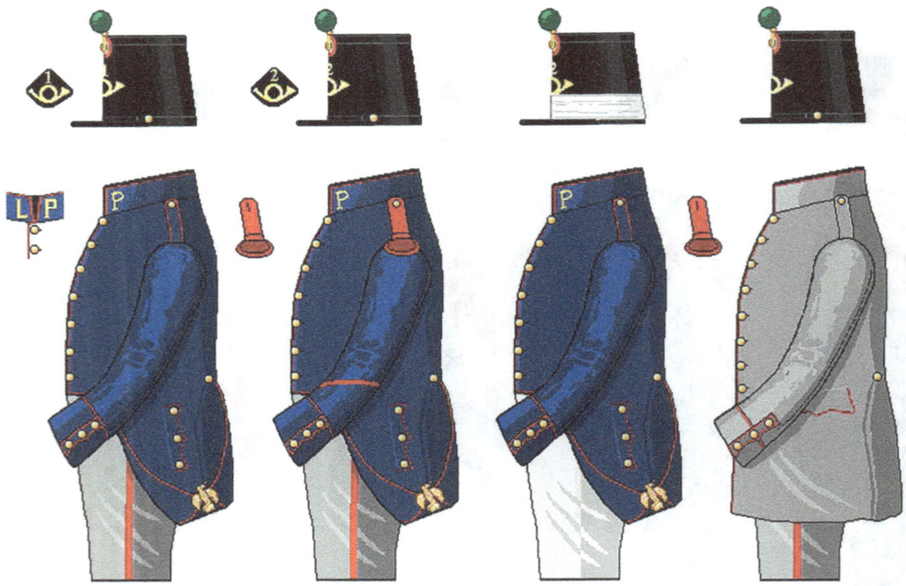

Private — Corporal
winter parade dress

Private
summer parade dress

Private
campaign dress with shako

First, Second and Third regiment

MEXICAN LIGHT INFANTRY

"Carabinero" winter parade dress — Private campaign dress
First, Second and Third regiment

Private — Bugler
Fourth regiment

MEXICAN LIGHT INFANTRY

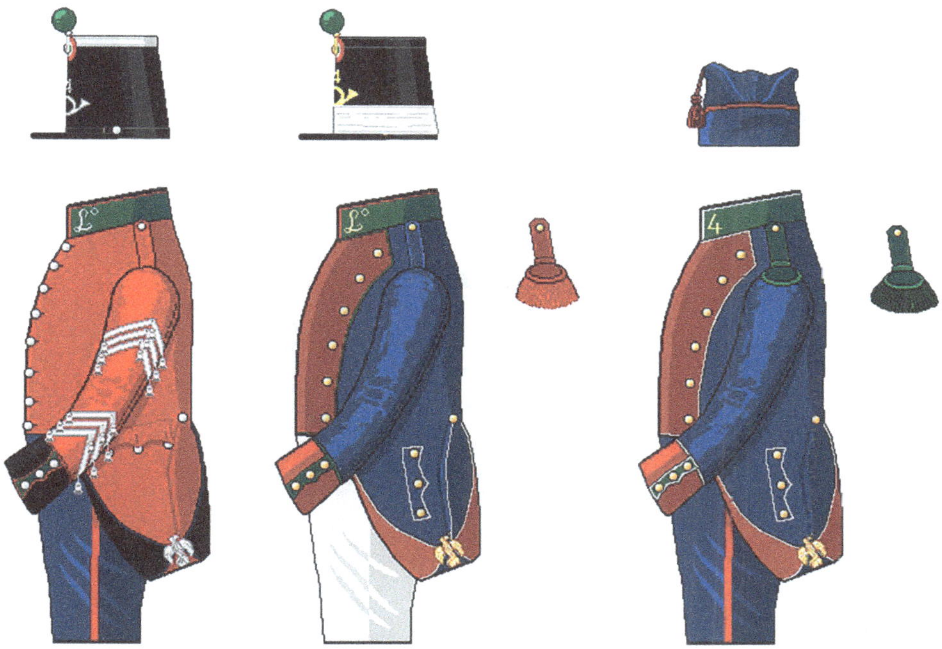

Trumpeter — Private summer dress — Private campaign dress

MEXICAN ACTIVE MILITIA

First Active Regiment
summer dress   winter dress

Second Active Regiment
early dress   late dress

MEXICAN ACTIVE MILITIA

Toluca
grenadier

Guanajuato   Queretaro
infantrymen

MEXICAN ACTIVE MILITIA

Puebla
winter dress summer dress

Invalids of Puebla

Invalids of Tampico

MEXICAN ACTIVE MILITIA

Active Militia
winter dress summer dress

Coast Guard Militia
Acapulco Tampico

MEXICAN ACTIVE MILITIA

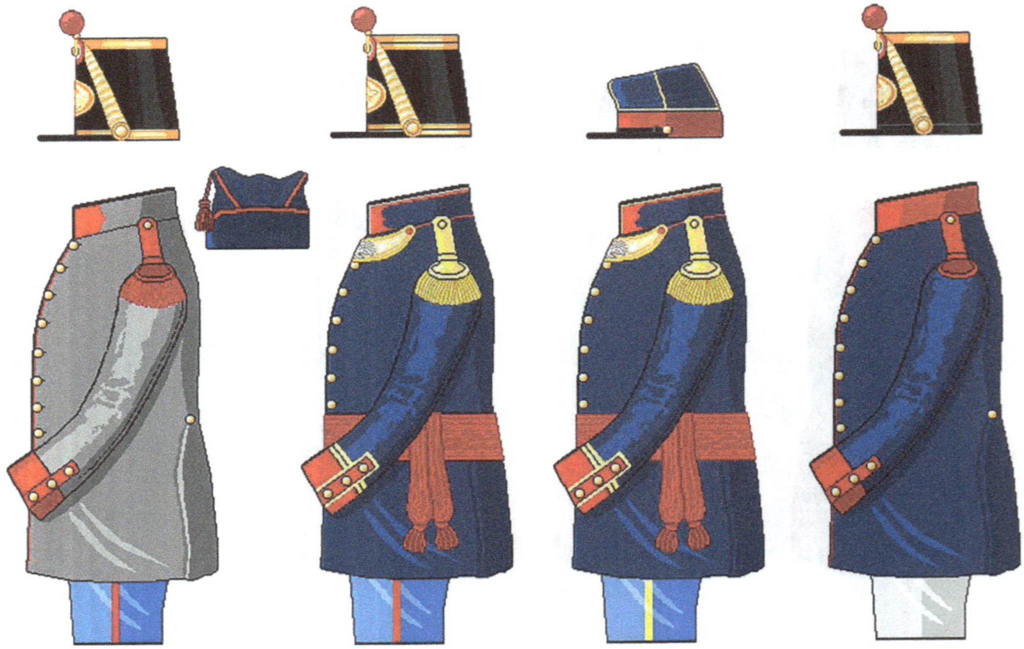

Coast Guard Battalion of San Blas
Private Captain

National Guard
of Angostura

Cavalry

The cavalry of the Mexican Army was organized on 24 December 1821, by melting together the many mounted units that were included into Iturbide's "Ejército Trigarante". A large number of the latter were not former "fijas" units but semi-regular ones made up of insurgents; as a result, when the cavalry of the new Mexican Army was organized, only the best elements of the insurgent units were retained in service and all the others were not admitted into the regular forces. Initially the Mexican cavalry did comprise eleven regiments, augmented to thirteen during 1822 with each regiment given a distinctive progressive number. In 1826 the 5th Cavalry Regiment mutinied and was thus disbanded for insubordination; as a result, the 13th Cavalry Regiment became the new 5th Regiment and the whole number of Mexican cavalry units was reduced to twelve (the 13th Regiment was not reformed).

Each regiment was comprised of four squadrons and each squadron was composed of two troops. A single troop had the following establishment: one captain, one lieutenant, two cornets, four sergeants, six corporals, two trumpeters and fifty-six troopers. As a result, a single squadron comprised a total of one hundred forty-five men (including a commandant). The staff of a cavalry regiment had the following composition: one colonel (commanding the unit), one lieutenant-colonel, two squadron commandants, one captain as major, two lieutenants, one chaplain, one surgeon, one farrier with two adjutants, one armourer, one trumpet-major, one corporal and eight "zapadores" (sappers). The zapadores had the same functions of their infantry equivalents and thus acted as a sort of "combat engineers". All the regiments were "line" ones, but each single unit was able to perform different tactical functions. The cavalrymen of each 1st Squadron were trained as light horsemen; those of each 2nd Squadron and 3rd Squadron were trained as heavy cavalry; those of each 4th Squadron were trained as medium cavalry. In addition each cavalry troop did comprise twelve horsemen armed with lances, for a theoretical strength of ninety-six lancers for each regiment (who could be grouped in a single "lancer squadron" in case of need).

On 19 November 1833 the existing twelve cavalry regiments were merged in order to form six consolidated regiments; these had the same internal organization of the previous ones but received new distinctive names instead of the previous progressive numbers. Differently from what happened for the infantry, the new designations were taken from notable episodes of the Independence War fought against the Spaniards: Dolores, Iguala, Cuautla, Vera Cruz, Palmar and Tampico. During the following years the general establishment of the Mexican cavalry was expanded with the creation of some new units: two regiments were added on 16 March 1839 and another one was raised during 1841. The new regiments were formed as a result of the experiences made by the Mexicans during the Texan Revolution: the territory of Mexico was extremely large and thus it needed a considerable number of highly-mobile troops in order to be properly patrolled. According to the 1839 reorganization, this was the internal composition of a cavalry troop/company: one captain, one lieutenant, two ensigns, one sergeant 1st class, four sergeants 2nd class, nine corporals, two trumpeters and fifty-two troopers. The staff of a cavalry regiment was to comprise the following elements: one colonel (commanding the unit), one lieutenant-colonel, two squadron commandants, four lieutenants, four standard-bearers, one chaplain, one surgeon, one corporal and twelve "zapadores" (sappers). On 28 April 1835 the first and only regular light cavalry regiment of the Mexican Army was raised, known as "Regimento Ligero de Mexico". This had the same internal structure of a line regiment but lighter equipment. Until 1835 the Mexican cavalry had comprised only three small independent light units: the "Yucatan Squadron" raised on 16 October 1826 and disbanded on 1 December 1847; the "Tabasco Permanent Cavalry Corps" raised on 16 October 1829 and later merged with an Active Militia company from Chiapas in order to form an independent squadron (which was disbanded on 1 December 1847); the "Isla del Carmen Permanent Cavalry Company" raised on 16 October 1829 and disbanded after a few years. On 20 October 1841 the independent "Puebla Light Cavalry Squadron" was also formed, which was disbanded on 1 December 1847. On 31 August 1847, due to the heavy losses suffered during the war with the United States, the "Regimento Ligero de Mexico" was melted with the remains of some Active Militia cavalry units in order to form the new "Vera Cruz Light Cavalry Regiment".

MEXICAN LINE CAVALRY

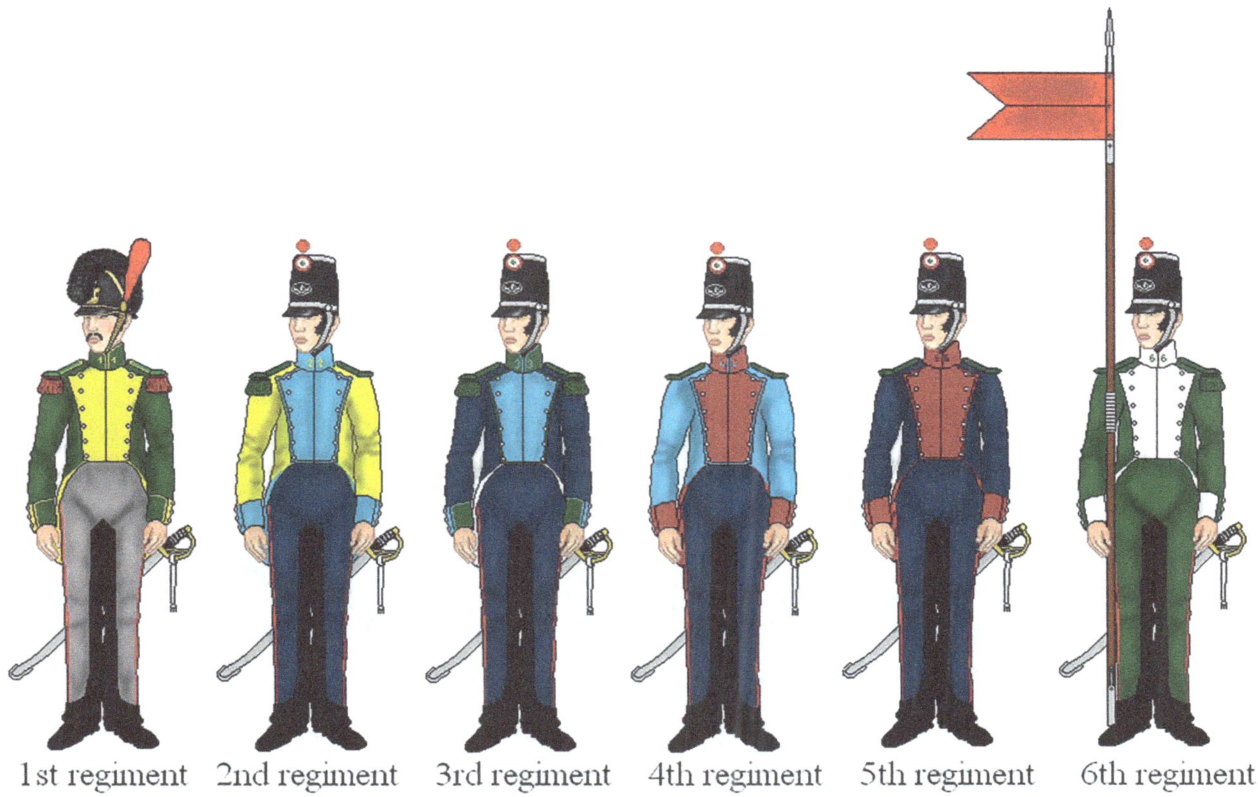

MEXICAN CAVALRY

7th regiment | 8th regiment | 9th regiment | Light Regiment of Mexico | Tulacingo Cuirassiers | Jalisco Lancers

MEXICAN LINE CAVALRY

Captain | Soldiers 3rd regiment | Sergeant

Santa Anna had a particular predilection for his cavalry units and always gave them some flamboyant uniforms strongly influenced by the contemporary European military fashions. He always sponsored the creation of "special" units having the same denominations and equipment of the cavalry regiments deployed by the European armies; as a result, during the last years of his rule, some new cavalry corps with "special" uniforms were formed: cuirassiers, lancers and mounted chasseurs. On 12 June 1840 a "Tulacingo Active Militia Cavalry Squadron" was formed; this soon proved to be an excellent mounted unit and thus Santa Anna decided to transform it into a cuirassier corps. The squadron received the new denomination of "Tulacingo Cuirassiers" and was given cuirassier heavy equipment on 15 January 1842. Some months later, on 1 September 1843, it was made a permanent unit and was absorbed into the regular cavalry. During the war with the United States the Mexican cuirassiers fought with great valour on several occasions until being disbanded on 1 December 1847. The lancers were originally raised as two squadrons of Active Militia cavalry from Jalisco, on 19 July 1843. Simply known as "Jalisco Lancers", they fought with great determination against the Americans and were disbanded on 1 December 1847. The cavalry of the Active Militia did comprise also another lancer squadron, since 20 August 1823: this was the so-called "Vera Cruz Squadron", which was one of the units merged with the "Regimento Ligero de Mexico" to form the new "Vera Cruz Light Cavalry Regiment" (on 31 August 1847). The mounted chasseurs were initially raised as an Active Militia cavalry regiment, on 12 June 1840; some time later, on 20 September 1843, they received the new denomination of "Cazadores a Caballo" and new light cavalry uniforms. In December 1847 the surviving elements of the unit were absorbed into the "Puebla Light Cavalry Squadron" shortly before the latter was disbanded. Differently from the cuirassiers, the lancers and mounted chasseurs always remained Active Militia units.

Since 1823 the cavalry of the Mexican Active Militia did comprise eight regiments plus several independent squadrons and companies; all these units, like the infantry ones, had a distinctive "local" nature and were usually named after the city/province where they were garrisoned. By 1845 the mounted units of the Active Militia did comprise five regiments plus seventeen independent squadrons. The regiments were from Guanajuato, Morelia, Oaxaca, Queretaro and San Luis de Potosi; the squadrons were from Bravos, California, Chalchicomula, Chiapas, Chinahuapan, Cuernavaca, Durango (two squadrons), Ixtlahuaca, Jalapa, New Mexico, Orizaba, Sierra Gorda (two squadrons), Teloloapan, Tlaxcala and Tula. The Mexican cavalry comprised also another two categories of troops: the mounted units of the Coast Guard Militia and the so-called "Presidios" (Presidial Cavalry Companies). The formers consisted of four squadrons and six companies, which were stationed along the Mexican coastline and performed the same duties of their infantry equivalents. The Presidial Cavalry Companies, instead, were part of a defensive system that was inherited from the Spanish colonial period. The northern border of Mexico, stretching from the Atlantic Ocean in Texas to the Pacific Ocean in California, was immense and scarcely populated; in addition, it was inhabited by several native tribes of fierce warriors like the Apaches or Comanches. As a result of the above situation, the Spanish colonial authorities had started to garrison these large areas with small companies of light cavalrymen mostly equipped with lances; these, thanks to their great mobility, were able to move very rapidly on a vast territory and could contrast in a very effective way the raids of the natives. These "frontier soldiers" were deployed in small forts located along the most isolated places of the border, known as "presidios". The latters worked also as administrative centers for the Mexican government and usually comprised also a Catholic mission.

PRESIDIAL CAVALRY COMPANIES

Standard uniform   Lower California   Santa Fe   Upper California

The cavalrymen garrisoning the "presidios" were commonly known as "cuera" (i.e. "leather") soldiers, since they usually wore protective leather coats over their uniforms (in order to have some better defence against native arrows and spears). In addition, sometimes the cavalrymen of the Presidial Companies could also carry a small leather shield, a piece of equipment that would have been of no use in any other area of Mexico. The main tasks of the "cuera" soldiers included patrolling the border and launching punitive expeditions against the rebel native tribes. South of the defensive line made up by the "presidios", the Mexican government deployed some other presidial cavalry companies known as "volantes" (i.e. "flying"); these did not wear leather coats and thus were even lighter than those of the "cuera" soldiers. In case a band of native warriors would have broken the defensive line of the "presidios", it would have been followed in close pursuit and later intercepted by the "volante" companies. In 1826 there was a Presidial Company in each of the following locations: Monclava, Agua Verde, Bahia, Rio Grande, Bahia del Espiritu Santo, Béjar, Los Alamos and Lampazo in the territory of Texas/Coahuila; Ciudad Chihuahua, San Buenaventura, Janos, San Elzeario, El Norte, Carrizal and Principe in the territory of Chihuahua; Frontera, Tucson, Altar, Santa Cruz, Buena Vista and Pitic in the territory of Sonora/Sinaloa; Santa Fe, Bado and Taos in the territory of New Mexico; San Francisco, Monterrey, Santa Barbara, San Diego, Fronteras and Loreto in the territory of the Californias. Finally, there were also the two "volante" companies stationed at Tamaulipas (in a key location placed between northern and central Mexico). In January 1842 two new Presidial Companies were established along the important road that connected Mexico City with the port of Vera Cruz, in order to watch over some prisoners who had been detailed to perform road construction tasks. During the period taken into account, the central government also tried to form some local Auxiliary Cavalry Companies that would have been commanded by the regional governors in time of peace and that would have passed under the jurisdiction of the regular army in time of war. Due to the frequent abuses of the regional governors, however, the central government finally decided to abolish these auxiliary companies (which were never particularly effective). During the war with the United States, however, several thousands of irregular Mexican lancers coming from the great ranches of the country supported the regular army. These irregulars were "vaqueros", the Mexican equivalent of the American "cowboys" and thus were skilled horsemen who used to fight with lances and "lazos".

MEXICAN LINE CAVALRY

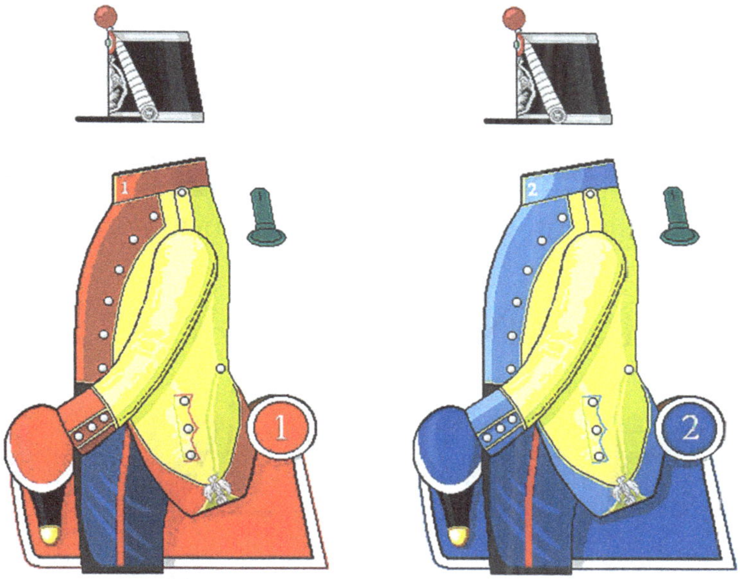

First regiment Second regiment

MEXICAN LINE CAVALRY

Third regiment

Fourth regiment

MEXICAN LINE CAVALRY

Fourth regiment

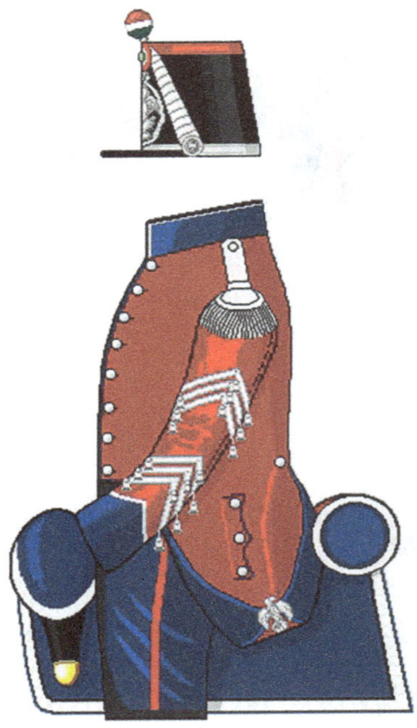

Trumpeter

MEXICAN LINE CAVALRY

Fifth regiment

Sixth regiment

MEXICAN LINE CAVALRY

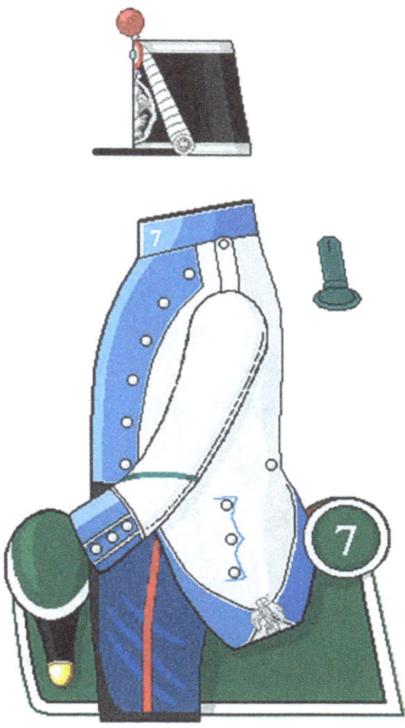

Seventh regiment

Eighth regiment

MEXICAN LINE CAVALRY

Ninth regiment

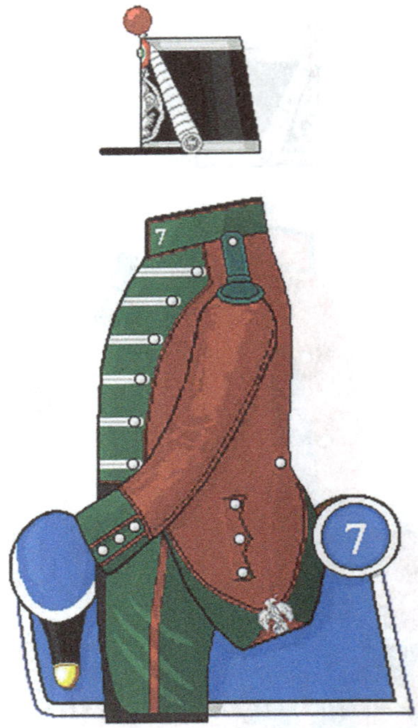

Seventh regiment
new uniform

MEXICAN LINE CAVALRY

First regiment
new uniform

New "unified"
cavalry uniform

MEXICAN LIGHT CAVALRY

Light regiment of Mexico

Light squadron of Puebla

TULACINGO CUIRASSIERS

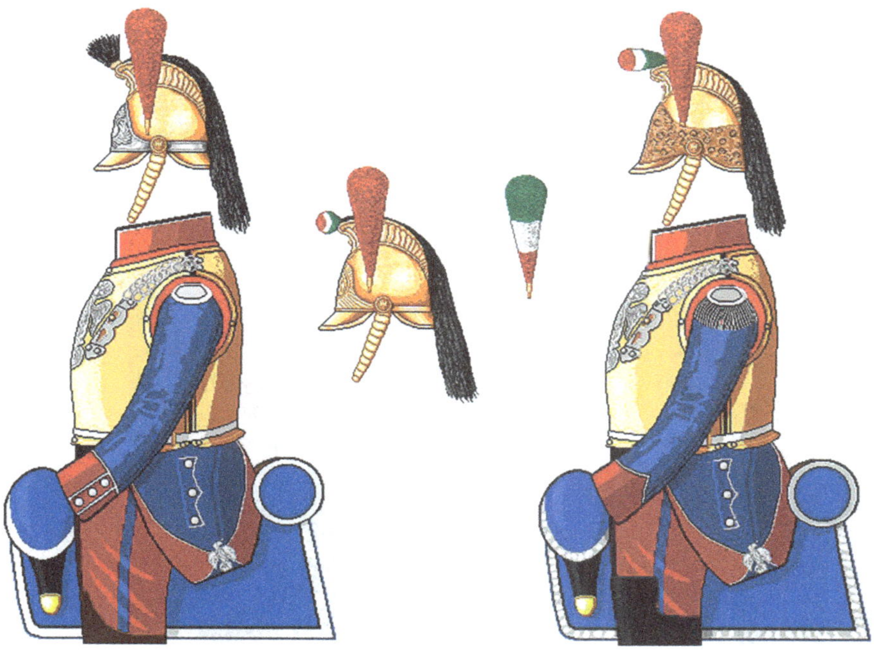

Trooper

Officer

TULACINGO CUIRASSIERS

Trooper

Officer

service dress

JALISCO LANCERS

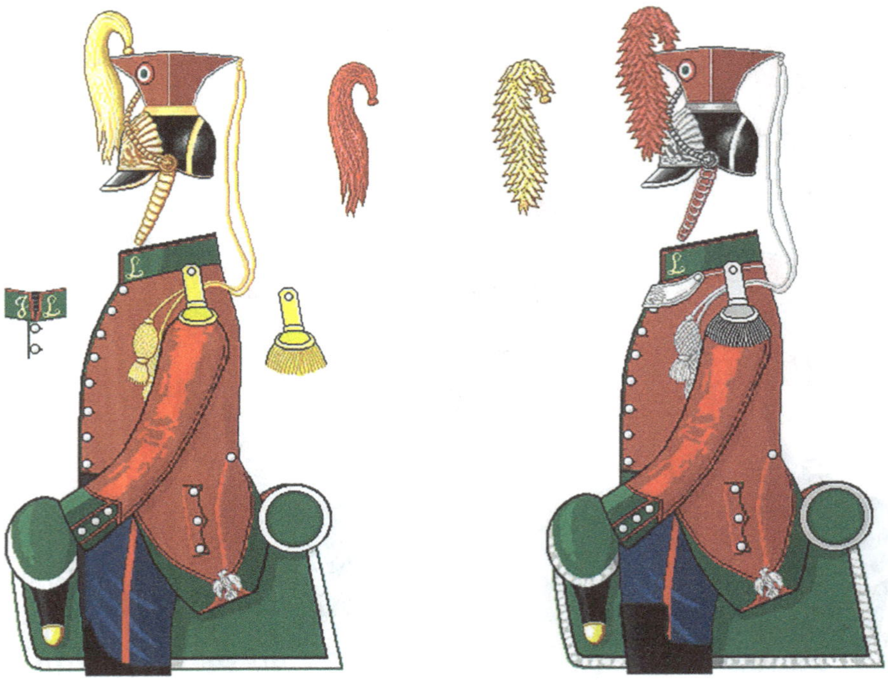

Trooper

Officer

MOUNTED CHASSEURS

Troopers

Officer

ACTIVE MILITIA CAVALRY

Parade dress

with helmet

with shako

ACTIVE MILITIA CAVALRY

Queretaro

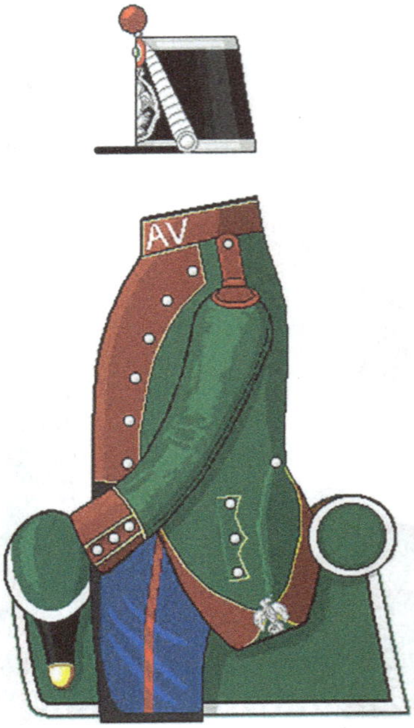

Lancers of Vera Cruz

ACTIVE MILITIA CAVALRY

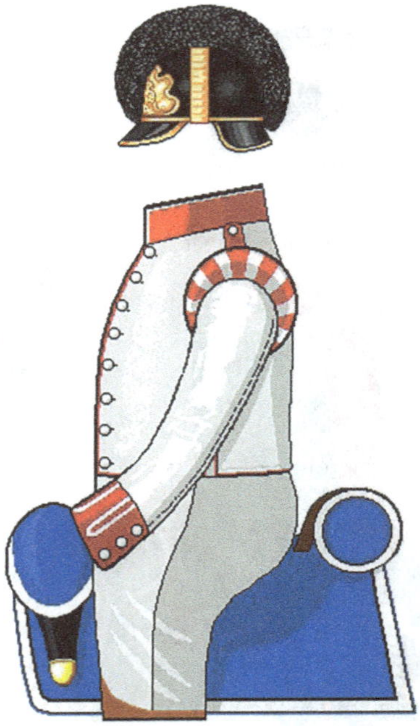

Tulacingo

winter dress

summer dress

ACTIVE MILITIA CAVALRY

Chihuahua

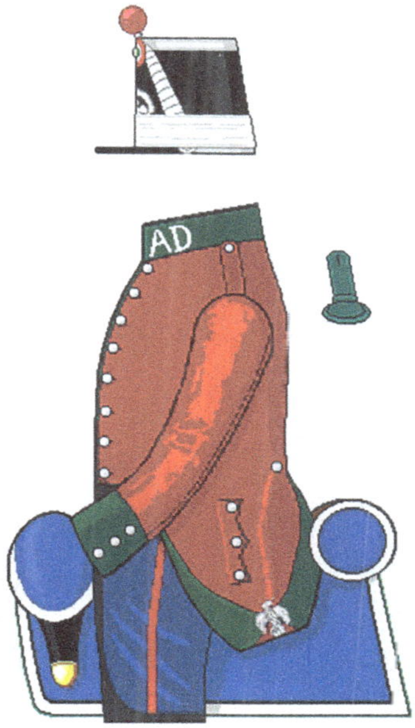

Durango

ACTIVE MILITIA CAVALRY

Zacatecas

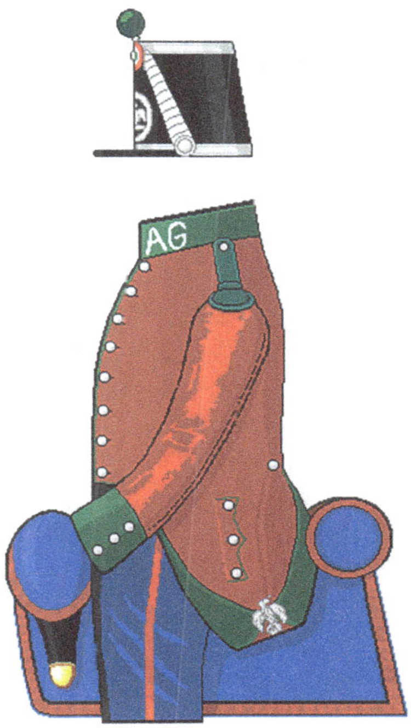

Guanajuato

PRESIDIAL CAVALRY COMPANIES

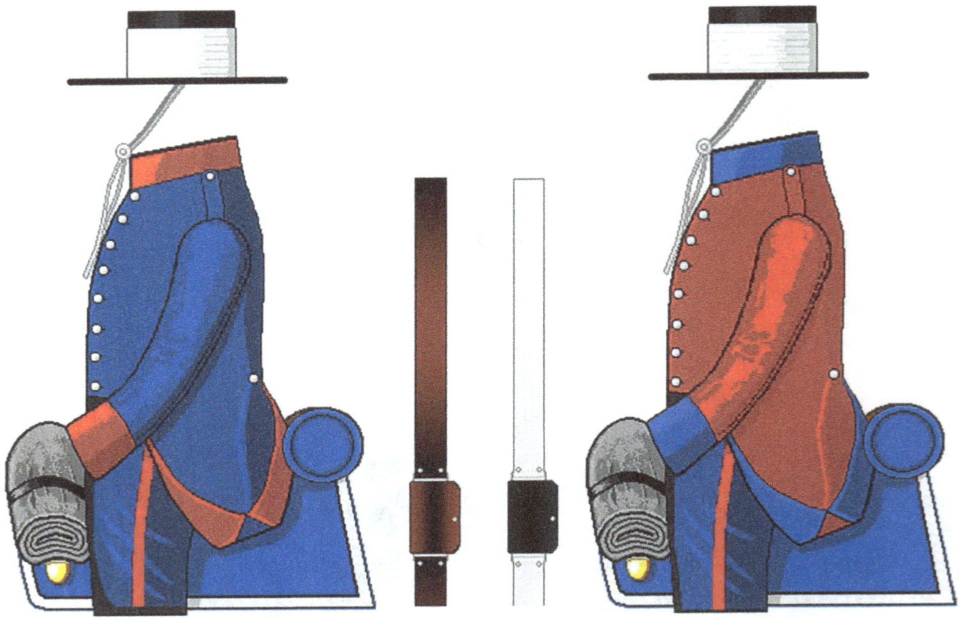

Texas and Coahuila

trooper

trumpeter

PRESIDIAL CAVALRY COMPANIES

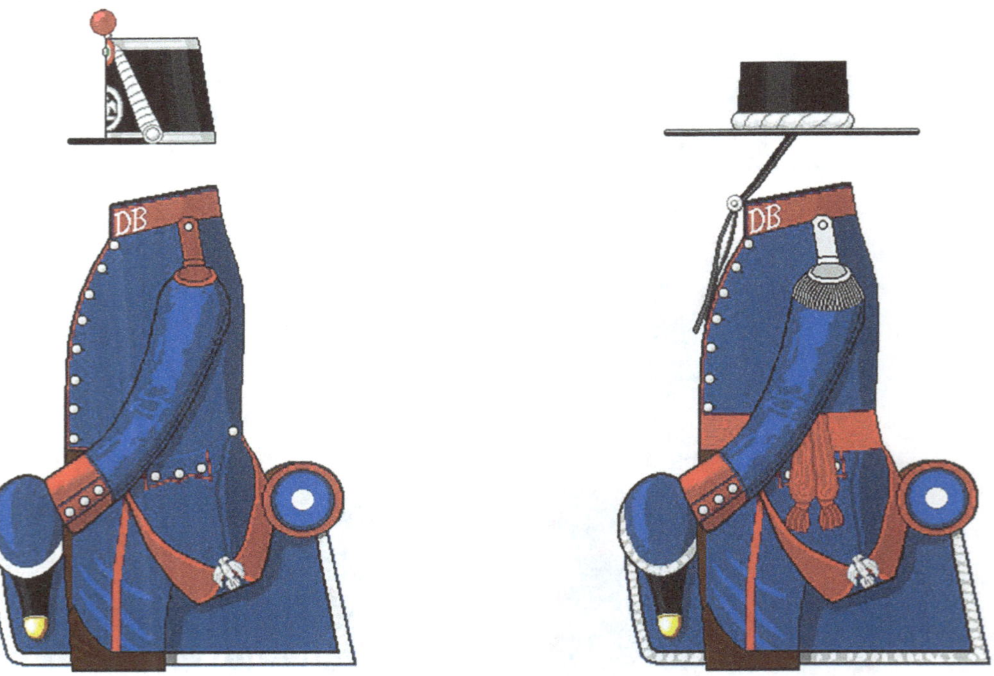

Bejar

Trooper

Officer

PRESIDIAL CAVALRY COMPANIES

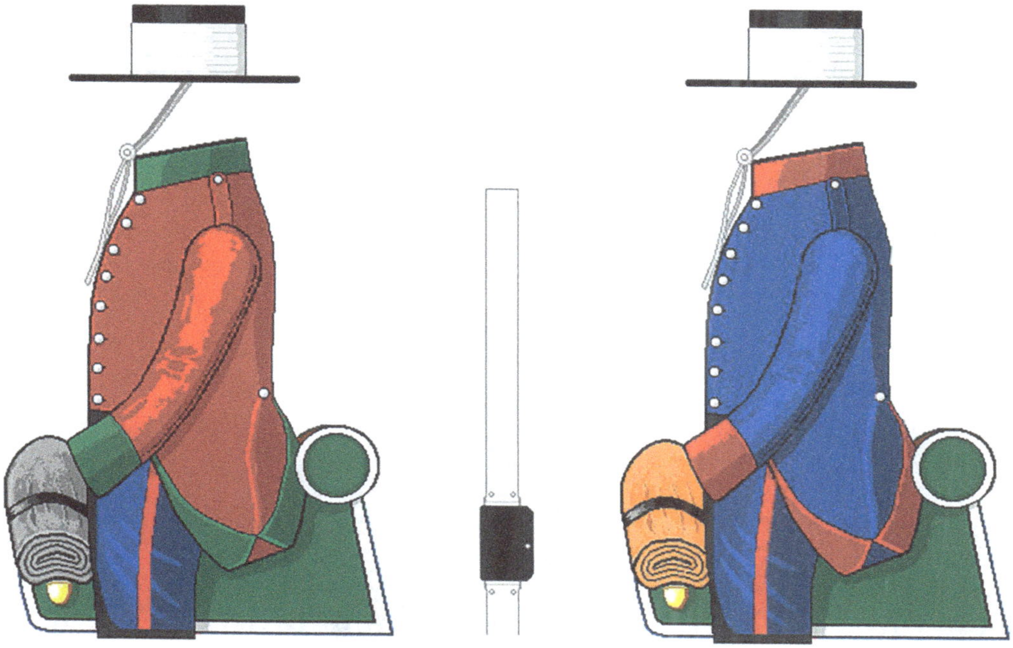

Chihuahua

PRESIDIAL CAVALRY COMPANIES

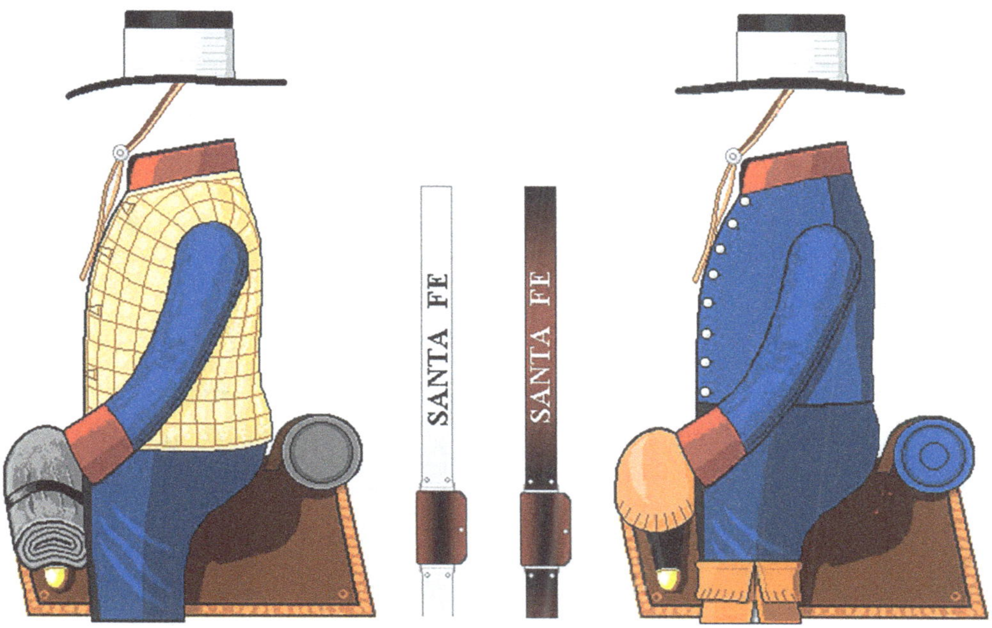

Santa Fe

PRESIDIAL CAVALRY COMPANIES

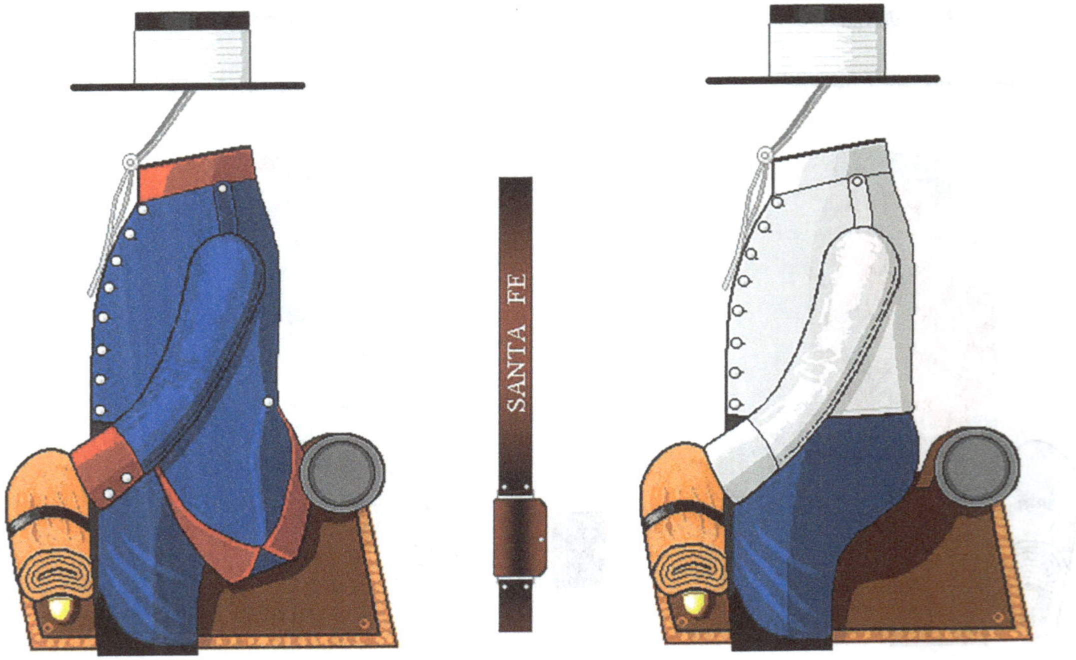

New Mexico

PRESIDIAL CAVALRY COMPANIES

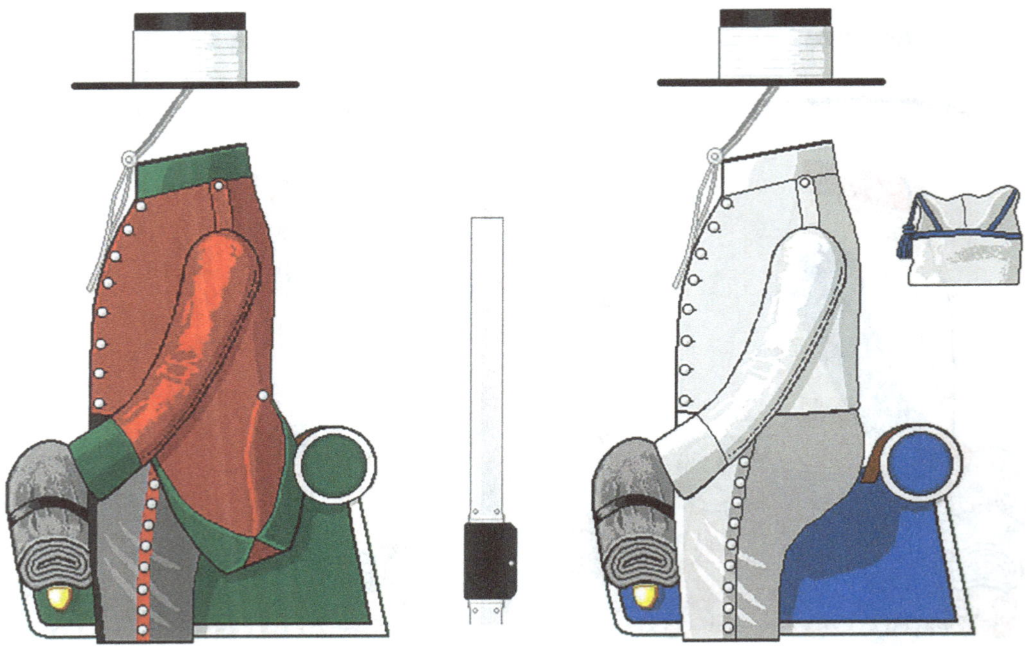

Lower California

PRESIDIAL CAVALRY COMPANIES

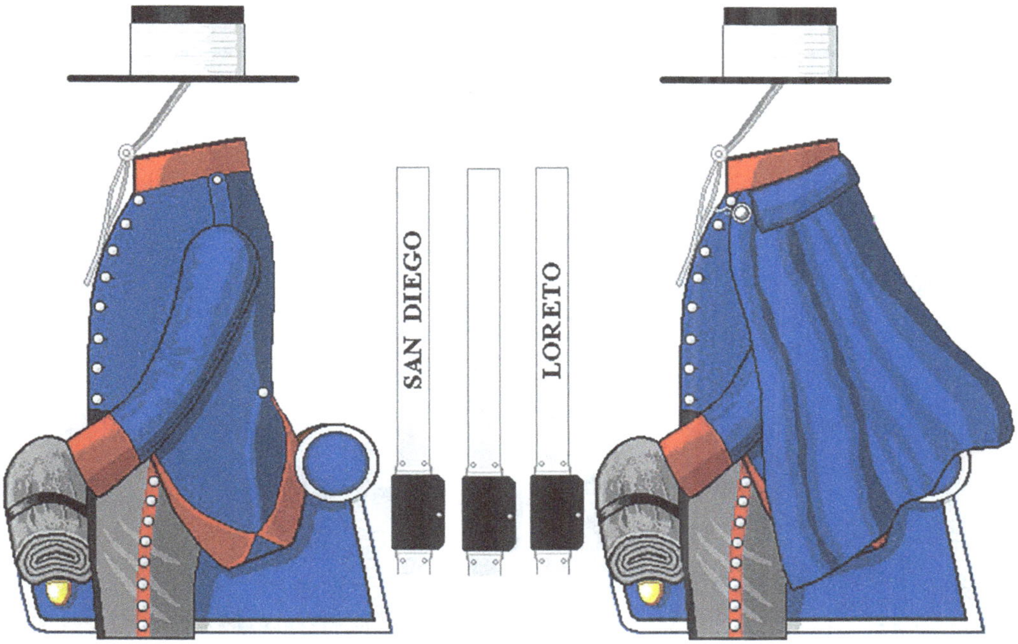

Upper California

PRESIDIAL CAVALRY COMPANIES

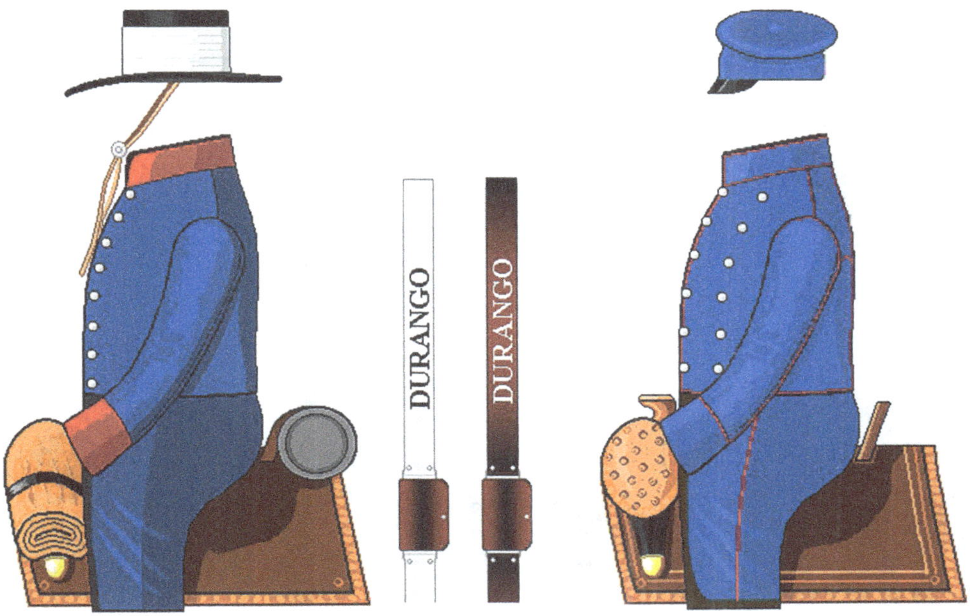

Durango

trooper officer

PRESIDIAL CAVALRY COMPANIES

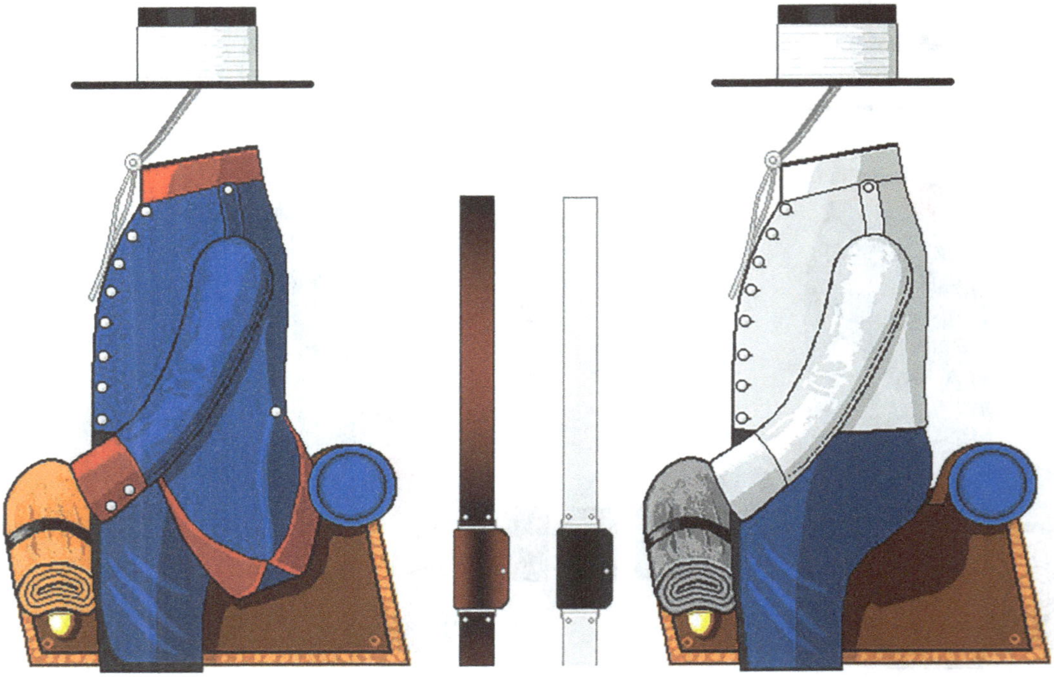

Sonora and Sinaloa

PRESIDIAL CAVALRY COMPANIES

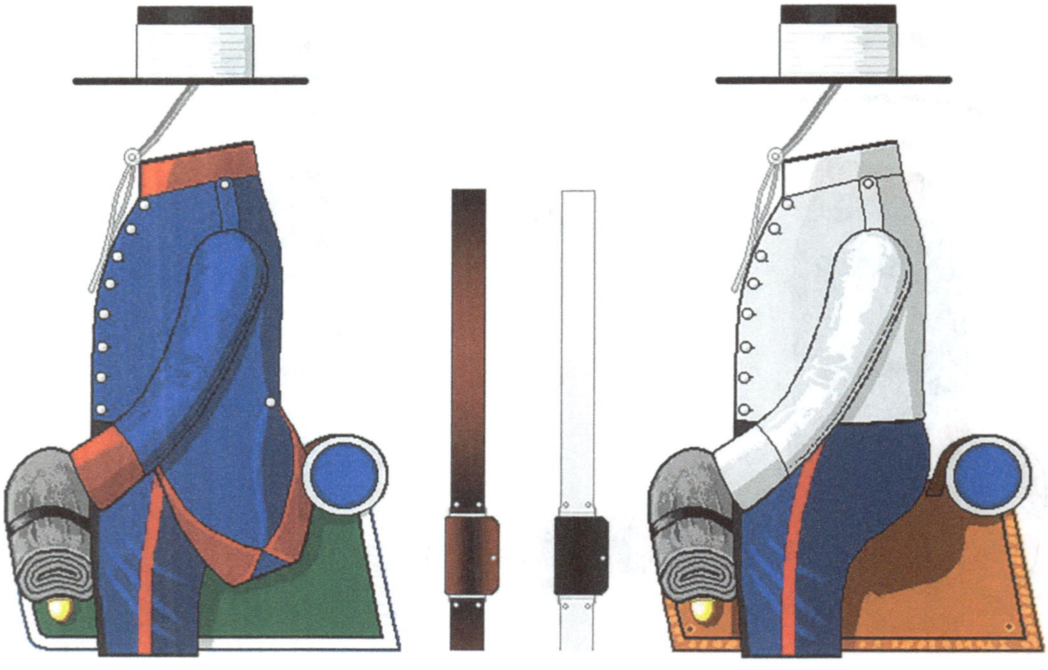

Flying companies of Tamaulipa

PRESIDIAL CAVALRY COMPANIES

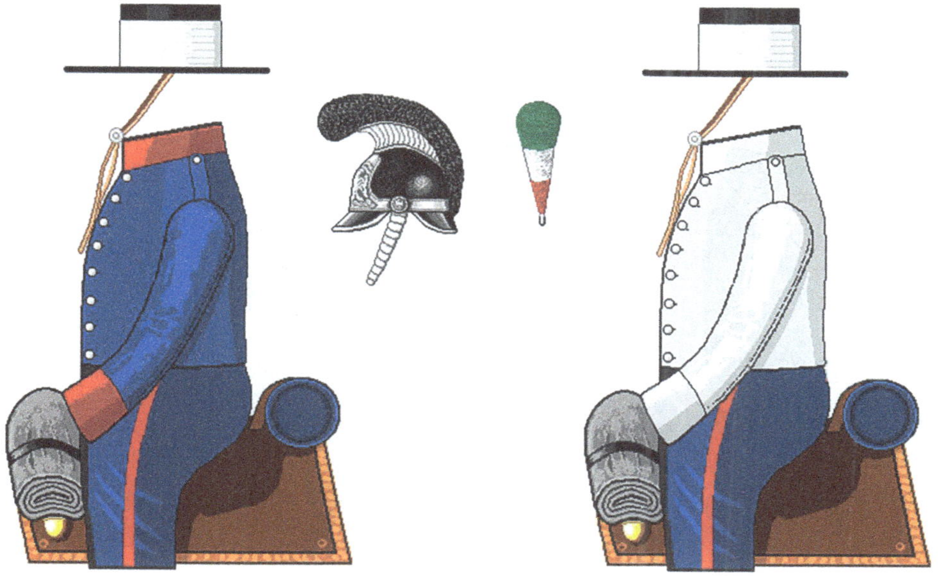

Presidial Companies of the Active Militia,
Northern Frontier

winter dress summer dress

PRESIDIAL CAVALRY COMPANIES

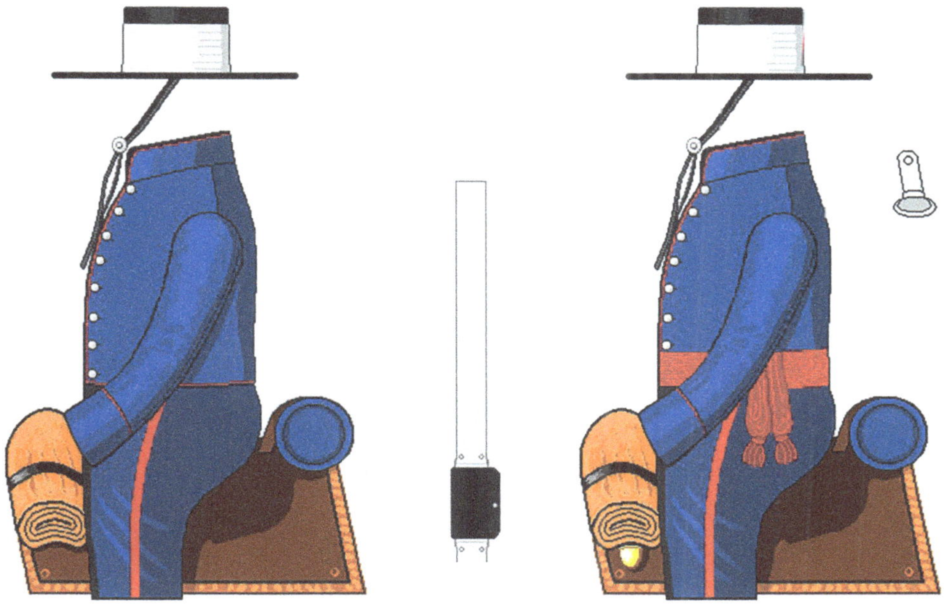

Presidial Companies of the Active Militia,
Northern Frontier

Trooper Officer

PRESIDIAL CAVALRY COMPANIES

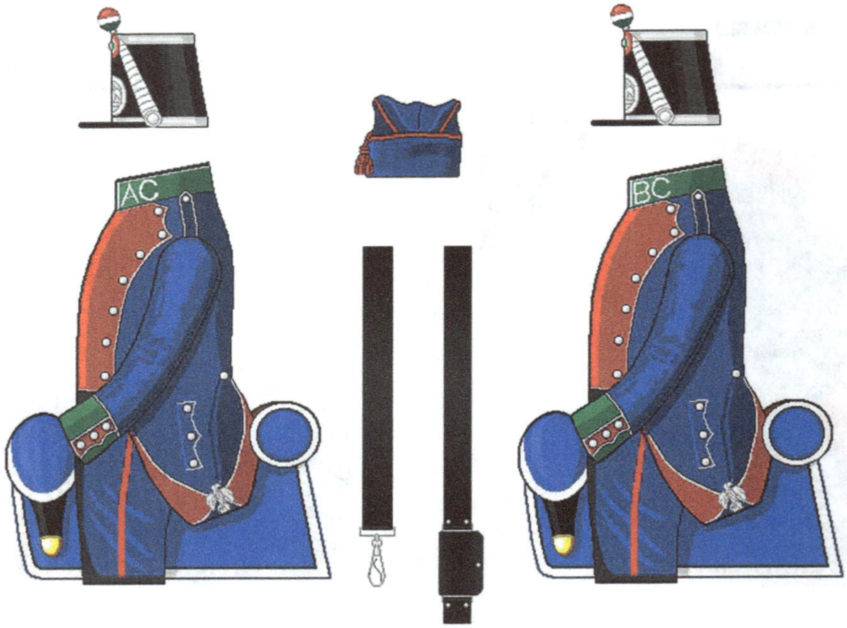

Upper California Lower California

parade dress

Artillery and Technical Corps

As all the Spanish colonies of Latin America, the Viceroyalty of New Spain did have a sizeable Royal Artillery Corps; this included a good number of well trained officers and of professional gunners who were mostly employed to garrison the many fortresses scattered across the territory of the colony. Not all the colonial artillerymen, however, manned fortress guns; there were, in fact, also several field batteries. When Mexico became an independent country, the artillery of the new Imperial Army comprised the following units: six foot companies, two horse companies and several fortress companies. On 14 February 1824 these were reorganized into a "National Artillery Corps" that was structured on three brigades. Two of the latters were of foot artillery, while the remaining one was of horse artillery. In total, there were 1,178 foot artillerymen and 589 horse artillerymen. The companies of fortress artillery, 11 in total with 1,152 gunners, were not part of the regular/permanent units but of the Active Militia (which mostly had garrison duties). In November 1833, due to economic reasons, the Horse Artillery Brigade was disbanded. After the combat experiences of the Texan Revolution, however, it became clear that the artillery would have played a fundamental role in case of a large-scale war with the United States. As a result, it became clear that whole National Artillery Corps had to be expanded and reorganized. By 1845 the Mexican artillery did comprise 266 officers and 4,989 gunners, assembled in the following units: three battalions of foot artillery, one battalion of horse artillery, nine

MEXICAN ARTILLERY

Captain Soldiers Sergeant

campaign dress

MEXICAN ARTILLERY

Parade dress
winter summer

Service dress
winter summer

MEXICAN ARTILLERY

Captain
parade dress service dress

Gunner
summer dress

Captain
campaign dress

MEXICAN ARTILLERY

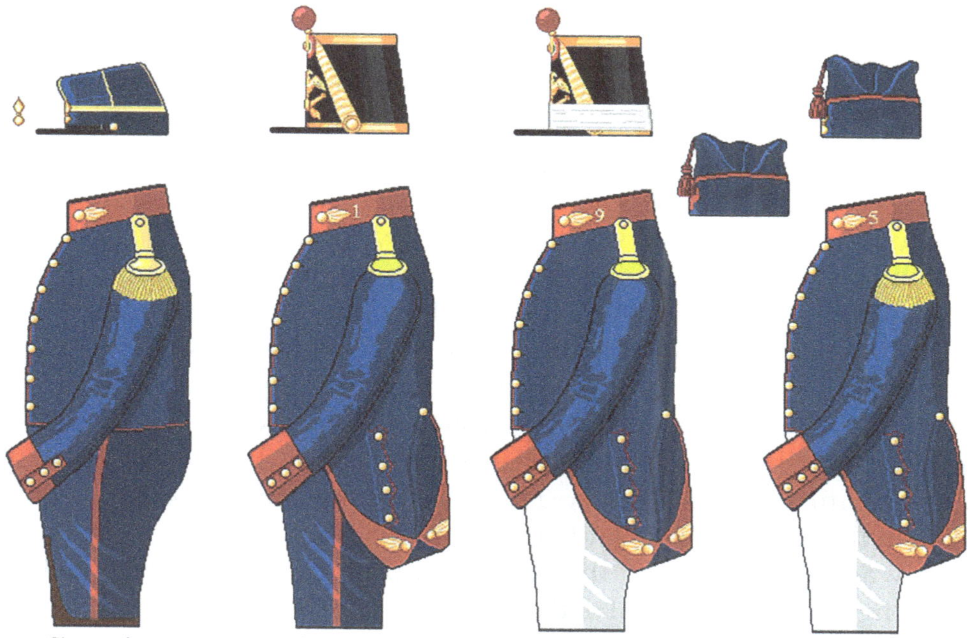

Captain Horse Artillery — winter dress, summer dress, Sergeant: Independent Garrison Companies

MEXICAN ARTILLERY

Gunner, Horse Artillery — Officer — Battalion of Sappers

independent companies of garrison artillery, three companies of artificers and two companies of the train. The four battalions of foot and horse artillery were also known as "brigades"; those of foot artillery had six batteries/companies of six guns each, while that of horse artillery had eight batteries/companies of six guns each. In total, the Mexican Army could deploy 120 field guns (48 of which were horse-drawn).

The staff of a foot artillery battalion comprised the following elements: five senior officers, six junior officers, one captain paymaster, one chaplain, one surgeon, one 1st class sergeant, one drum-major, one bugle-major, two bandmasters, twelve musicians, one armourer and nine pioneers (one corporal and eight soldiers). The internal composition of a foot artillery battery/company was as follows: three officers, two buglers, two drummers, 20 NCOs and 66 gunners. The staff of the horse artillery battalion comprised the following elements: five senior officers, six junior officers, one captain paymaster, one chaplain, one surgeon, one 1st class sergeant, one trumpet-major, one bugle-major, two bandmasters, twelve musicians, one armourer, one groom marshal, two saddlers and nine pioneers (one corporal and eight soldiers). The internal composition of a horse artillery battery/company was as follows: four officers, two trumpeters, 20 NCOs and 66 gunners. Each single mounted battery had 88 saddle horses and 50 draft horses. In addition to the above units, there were also a few officers who supervised the production of new artillery materials in the state arsenals and one company of artillery workers. The whole National Artillery Corps was commanded by the Artillery Corps Headquarters, which consisted of 25 senior officers having specific technical skills.

The Mexican Army's Corps of Engineers was first organized during March 1822, following the Spanish colonial model. It did comprise only a small number of officers, who had specific technical skills and who were mostly charged with the supervision of all the fortifications located on the national territory. By 1825 the Corps of Engineers consisted of one director, three lieutenant-colonels, four captains, two lieutenants and 18 sub-lieutenants. In time of war, the engineer officers performed a series of fundamental tasks: they established strongpoints, performed reconnaissance missions along rivers in order to discover fords and build bridges, repaired or built new roads, directed assaults against the fortified positions of the enemy, organized the defence of besieged forts/ports, planned and constructed any kind of fortification needed by the army. The engineer officers had a certain number of enlisted men under their direct command, who were formally organized into an independent unit known as Battalion of Sappers ("Zapadores"). This was first raised only in 1827 and was structured on six companies with 100 men each. The 1st Company of the battalion was of Miners, the 2nd Company was of Pontoniers and the other four were of proper Sappers. A standard company of "Zapadores" comprised the following elements: three officers, two drummers, one fifer, five NCOs and 78 sappers. Apparently the Battalion of Sappers was a real elite unit of the Mexican Army, since its members were highly disciplined and underwent some specific training. As a result, the battalion was mostly employed as an elite line infantry unit rather than a corps of combat engineers. This is confirmed by the fact that the survivors of the Grenadiers of the Guard were absorbed into the Battalion of Sappers when their own unit was formally disbanded. The "technical corps" of the Mexican Army did comprise also a small Corps of Marines; this was formally part of the Mexican Navy and not of the Mexican Army. It had been formed soon after Mexico's independence and was mostly charged with the task of garrisoning the main ports/naval installations of the country. During the period taken into account, the Corps of Marines consisted of seven companies of naval infantry and one company of naval artillery.

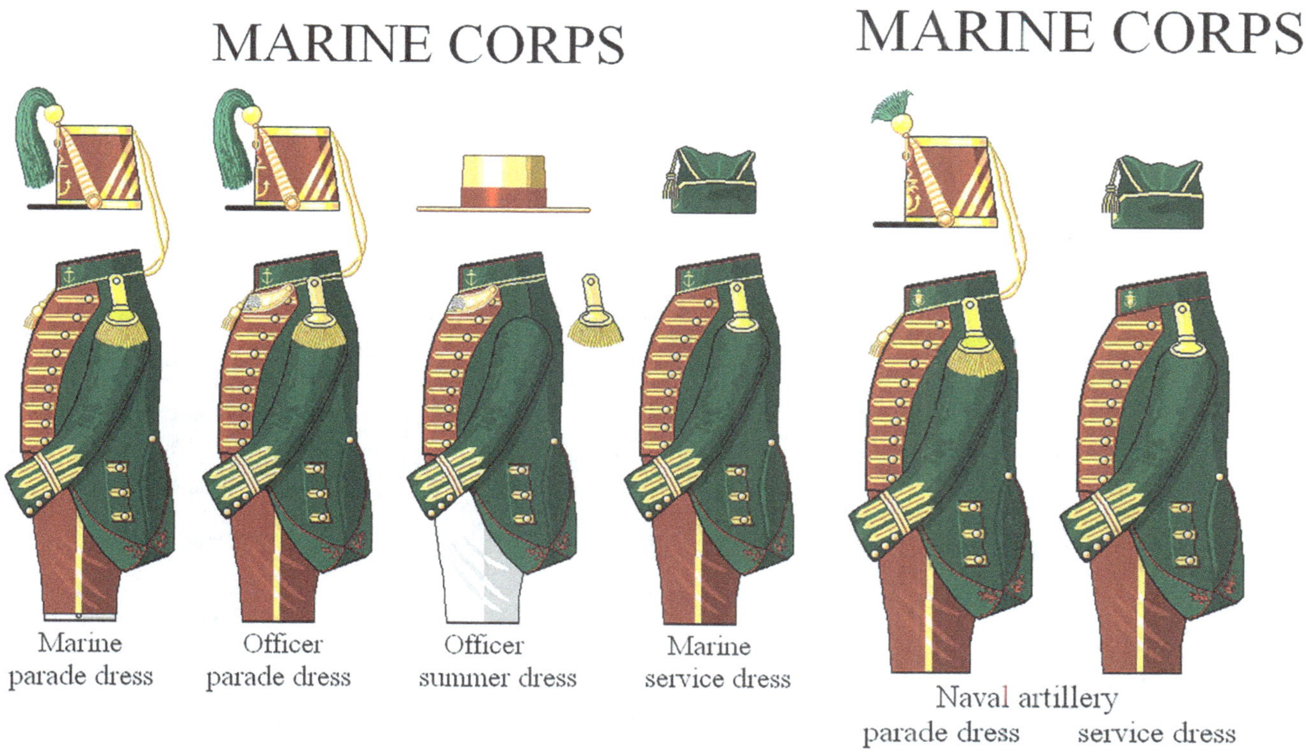

Among the "technical corps" of the Mexican Army there was also the Military College, commonly known as "Colegio Militar". This was created for the instruction of future officers in 1823 and later transferred to Chapultepec Castle (Mexico City) in 1841. Originally the Military College was to comprise only 100 "cadets" organized into one company, but in 1833 the general establishment of the "Colegio Militar" was increased to 200 students organized into two companies. During the following years, however, the actual number of cadets was always greatly inferior than the planned one; by the outbreak of the war with the United States, there were less than 50 young students in the Chapultepec Academy. The three-year curriculum of the latter comprised mathematics, chemistry, physics, tactics, artillery, fortification, architecture, astronomy and land surveying. From a formal point of view, the "Colegio Militar" was under control of the Engineer Corps and (at least on paper) it was to provide 40/50 new junior officers to the army every year. Of the two companies that made up the military school, one was known as "Cadet Company" and the other as "Sublieutenant Company". Each company had an infantry captain as commander and a cavalry captain as second in command. Theoretically, the two companies were divided into squads of eight cadets commanded by one Cadet Corporal. Two squads with 16 cadets and 2 corporals were put under command of a Cadet Sergeant. When the Americans attacked Chapultepec Castle on 13 September 1847, the few cadets who were garrisoning the fortification fought with great courage and six of them were killed. These are still remembered today among the greatest heroes of the Mexican nation. In December 1843 the Mexican government ordered the creation of a special Artillery School for the formation of the future artillery officers, which was to be a separate entity from the existing "Colegio Militar", but this never became reality due to the chronical shortage of funds and the outbreak of the war with the United States.

Chapultepec Castle as it looks today

Civic Militia, National Guard and Foreign Volunteers

The urban and sedentary militias inherited from the Spanish colonial government were reorganized by the Mexicans on 8 August 1822, when they were all assembled together in order to form the so-called Civic Militia. This was to comprise units of infantry, cavalry and artillery: the infantry would have been organized into single companies of 70-100 men or in battalions with 4-6 companies; the cavalry would have been organized into single troops with one captain, one lieutenant, one sub-lieutenant, two trumpeters and 60 troopers; the artillery companies, which were garrisoned in the major cities and ports of the various Mexican provinces, comprised three officers (one captain, one lieutenant and one sub-lieutenant) plus 40 gunners (including one drummer and one fifer). The main functions of the Civic Militia included capturing bandits, guarding public buildings, escorting public officials and transporting prisoners. Each unit bore the name of its home province/city, excepting those who came from the Mexican capital of Mexico City; the latter, like Washington DC in the United States, was not part of any Mexican state and thus formed an independent "Distrito Federal" (i.e. Federal District). Since December 1827 the Civic Militia came under direct control of the various state governors, who became directly responsible for its administration and for its minimum enlistment quotas (established by the central government according to the population of each Mexican state).

CIVIC MILITIA INFANTRY

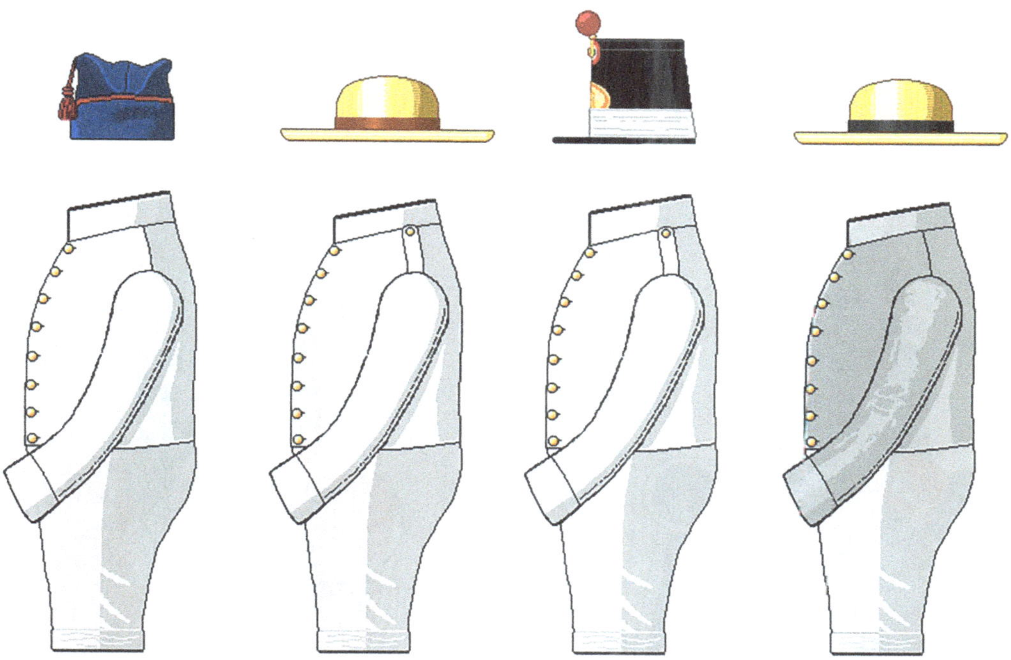

Variations of summer dress

CIVIC MILITIA CAVALRY

Standard uniform

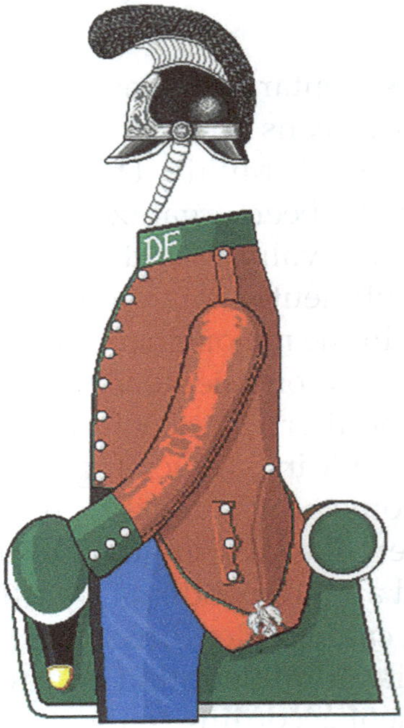

Federal District

CIVIC MILITIA CAVALRY

Texas and Coahuila

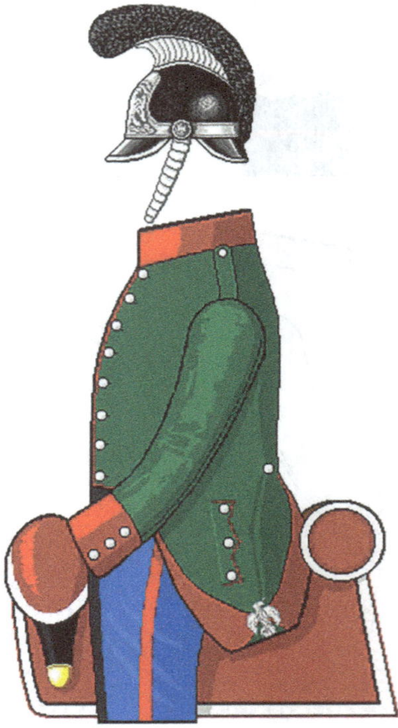

Standard uniform

CIVIC MILITIA CAVALRY

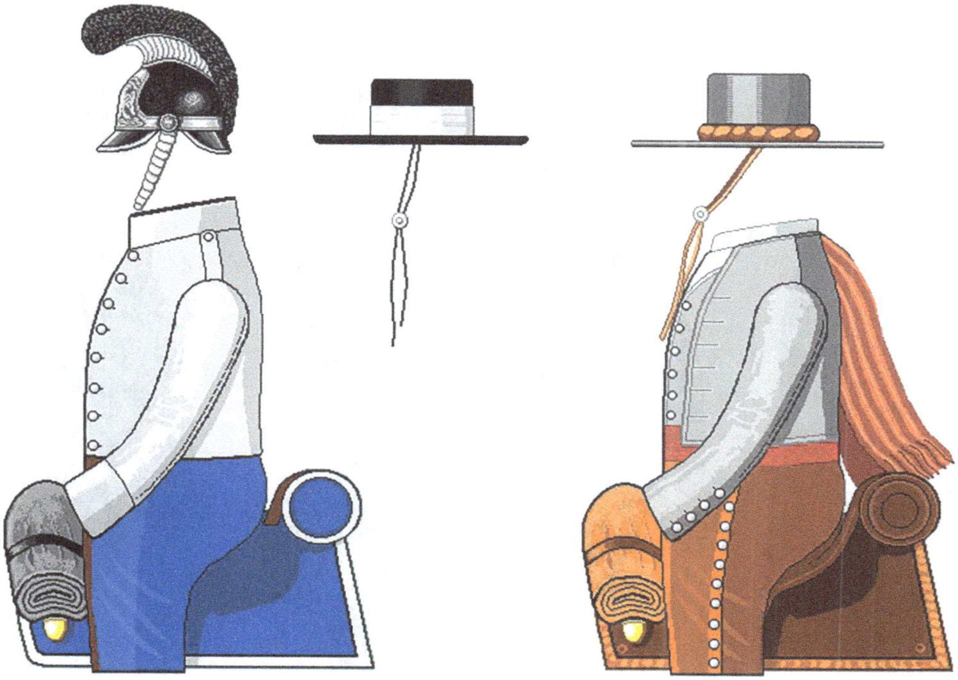

Summer dress

Civilian dress

CIVIC MILITIA ARTILLERY

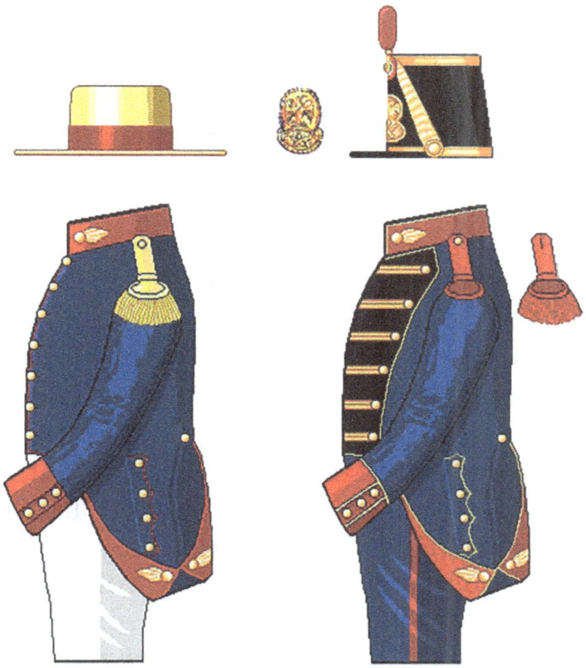

Officer

Gunner

MEXICAN FOOT TROOPS

Commerce Battalion winter dress — Commerce Battalion summer dress — Battalion of Invalids — Independent companies of regular infantry

Very soon the various governors, especially those who operated in the states that were located far from the Federal District, understood that they could employ the units of the Civic Militia in order to achieve their own personal goals. As a result, during the period taken into account, the members of the Civic Militia played an important role during several of the internal rebellions that shattered Mexico. Since it was possible to avoid service in the Civic Militia by paying the small sum of three "reales", usually only the poorest citizens and the ex-criminals were called to serve in its ranks. On 16 April 1833, in order to have a local force that could be strong enough to face the attacks coming from the Active Militia of other Mexican states, the Active Militia of the Federal District was completely reorganized. Now it was to comprise three infantry battalions, one squadron of cavalry and one brigade of artillery. The cavalry squadron was later disbanded in order to furnish the men needed to create the new "Hussars of the Guard of the Supreme Powers".

In addition to the above, there was also another "elite" infantry battalion known as "Commerce Battalion of the Federal City" (raised on 14 November 1833). This revived the tradition of a famous militia regiment that had existed in Mexico City under Spanish colonial rule, the "Commerce Militia Regiment". The new unit, like the old colonial one, had officers who came from the wealthiest citizens of the Mexican capital: these were mostly rich merchants or privileged professionals, who wished to serve in the Civic Militia in order to show their social status. NCOs and soldiers of the "Commerce Battalion" all had a good education or a decent trade. The unit served without pay and its duties were limited to maintaining public order for property or people in the urban area of Mexico City.

In 1845, shortly before Santa Anna's return to power, the Mexican government decided to abolish the Civic Militia in order to put an end to all the abuses committed by the state governors. Since Mexico had never had a true national volunteer force like the French National Guard, it was decided to replace the Civic Militia with a new organization known as "Guardia

Nacional". With this move the Mexican government hoped to create a large volunteer force that could have been of great use in the upcoming large-scale conflict against the United States. The new national guardsmen were not paid by the central government or by the authorities of their state; in addition, they had to be armed and clothed at their own expense. This meant that only the richest cities and provinces of Mexico were able to provide National Guard units having decent weapons and uniforms. According to the law that created the National Guard, these volunteers were to be called to serve only in case of war. From a formal point of view, the national guardsmen were placed under control of the state governors like the previous Civic Militia but they could be called to serve only by the central government. The National Guard of the Federal District did comprise four infantry battalions, which were organized in September 1846 and had distinctive names. According to a practice that was quite common in Latin America during the 19th century, the various National Guard units of Mexico City were made up by men who had the same profession in civil life. The "Victoria Battalion" comprised merchants and professionists; the "Hidalgo Battalion" was made up of clerks; the "Bravo Battalion" and "Independencia Battalion" were mostly made up of artisans. These four infantry units, together with the "Commerce Battalion" that was not disbanded, were commonly known as "polkos" (i.e. "polka dancers") because of their colourful uniforms. On 28 February 1847, instead of marching to Vera Cruz to help the local garrison that was attacked by the Americans, the National Guard battalions of Mexico City revolted against Santa Anna and tried to install a new government in the capital. In the end, after some fierce fighting, the regular troops sent to the capital by Santa Anna (including the Guard Hussars) were finally able to crush the rebellion and restore order in the Federal District. When the Americans launched their final assault on Mexico City, which ended the war, the remnants of the four National Guard battalions of the capital fought with great determination to defend their city and were completely annihilated.

To complete this general outline of all the Mexican Army's components, we have to analyse the history and organization of the famous "Saint Patrick's Legion". This was a unit of foreign volunteers, mostly deserters from the US Army, who decided to change side and fight under the Mexican flag. Since the beginning, the legacy of this peculiar military unit has been

MEXICAN NATIONAL GUARD

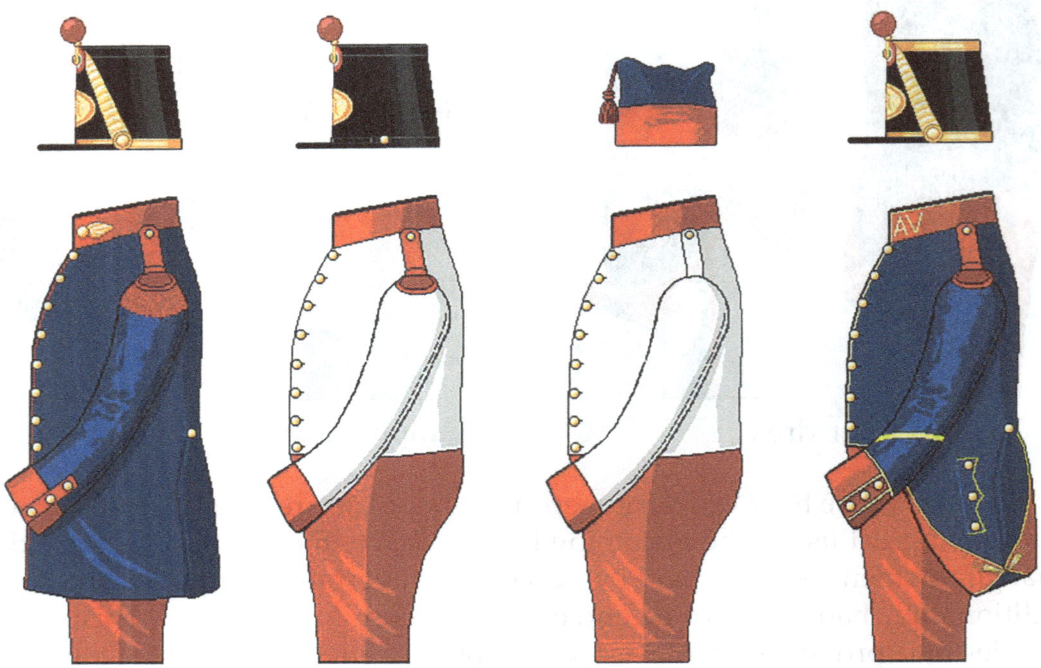

Battalions of Mexico City (Federal District)

Bravos Hidalgo Independencia Victoria

very controversial: for the government of the United States the members of "Saint Patrick's Legion" were simply traitors and renegades; for the Mexican people, the foreign volunteers of this unit who decided to fight for the freedom of their country were real heroes. In reality, the men of this "foreign legion" were both things: of course they were deserters but they were also valorous soldiers who served their new "motherland" with great determination.

In June 1847 the Mexican government decided to create a "Legion of Foreigners" by as-

SAINT PATRICK'S LEGION

Officer Soldier Corporal Soldiers campaign dress

sembling together all the foreigners who already fought in the ranks of the Mexican Army; some of these were recent immigrants coming from several European countries but the majority were deserters of the US Army. During the years before the outbreak of the war, thousands and thousands of Irishmen had abandoned their homeland due to the terrible "Great Irish Famine". These poor individuals were forced by the circumstances to cross the Atlantic Ocean, in what became known as the "Irish Diaspora". When these Irish families arrived in the USA, most of their able-bodied men had no choice but to enlist in the US Army in order to earn a living. Irishmen had always served as mercenaries in the most important European armies, most notably in those of Catholic powers like France and Spain; this time, however, the situation was completely different. First of all, the Irishmen of the US army were not professional mercenaries like their ancestors but desperate immigrants who needed an occupation; secondly, they were not serving under the flags of a Catholic country and thus had very little in common with the culture and religion of their new country. After the outbreak of the war with Mexico, malcontent began to spread among the Irish volunteers of the US Army: discipline was too harsh, rations were extremely poor and the general conditions of service were unsustainable. In addition, the American officers had a very negative opinion of their Irish recruits. As a result, many Irish soldiers of the US Army started to desert and join the ranks of the Mexican Army.

After all, Mexico was much more similar to Ireland than the USA: Catholicism was the only admitted religion and also the Mexican popular culture was quite similar to the Irish one. Very soon there were enough men to form the "Legion of Foreigners" cited above, whose mem-

SAINT PATRICK'S LEGION

Gunner Sergeant Officer
Active Militia Artillery Uniform

bers soon became known as "los colorados" because of their distinctive red hair. The Legion was organized as a single artillery company and comprised a total of 48 Irishmen. The Mexicans decided to employ the foreigners as artillerymen because they were among the few soldiers of the Mexican Army who did know how to use heavy artillery guns in an effective way. The company/battery of the "Legion of Foreigners", in fact, was initially equipped with three heavy guns. The unit was guided by the Irishman John Riley, who had already served as a NCO in the British Army of Canada and had later joined the ranks of the US Army. Riley's company fought with great distinction on several occasions and thus received many official awards from the Mexican government. The Americans, instead, started to hate this unit of "renegades" that caused so many casualties to them during several battles. Understanding that the foreign soldiers were much more effective than any other of its units, the Mexican Army decided to enlist as many deserters/mercenaries as possible in order to augment the numerical consistency of the Legion. To achieve this, the Mexican government promised great incentives to all those foreigners who would enlist in its army; these included the grant of citizenship, the payment of higher wages (superior to those of the US Army) and generous land grants. As a result of the above measures, an increasing number of Europeans living in Mexico enlisted into Riley's unit and more Irishmen deserted from the US Army in order to obtain a "ranch" in Mexico once the war was over. With the new recruits, it was possible to form a second "foreign" company and thus the unit became a battalion. This received the new denomination of "Saint Patrick's Legion" or

"Saint Patrick's Battalion". Formally this was to be an infantry unit, but Riley's men continued to retain and man their battery of heavy guns. The two companies had the same internal organization of the regular infantry ones but wore the same uniform of the Active Militia infantry; from a formal point of view, in fact, the "Saint Patrick's Legion" was an Active Militia corps. The battalion carried the famous "Green Harp Flag", which comprised the "Golden Harp of Erin" and the famous "Shamrock" symbol over a green background. This flag also bore the Gaelic inscription "Erin go Brach", which means "Ireland Forever". Under many points of view, the "Saint Patrick's Battalion" was the only Mexican response to the highly-mobile horse artillery units deployed by the US Army; the foreigners under Riley's command, in fact, were the only soldiers of the Mexican Army who could move heavy guns on the battlefield with the same speed of the Americans. On 20 August 1847, at the Battle of Churubusco, the men of the "Saint Patrick's Battalion" fought with great courage and caused heavy losses to the attacking Americans; after the clash, however, most of them were captured by the enemy. These, 50 in total, were hanged as traitors and deserters after a rapid trial during September 1847.

SAINT PATRICK'S LEGION

Soldier Sergeant Soldier Officer

Active Militia Infantry Uniform

SAINT PATRICK'S LEGION

Service dress Officer

Campaign dress Soldier

Summer dress

Mexican "rancheros", with their traditional dress

Mexican "vaquero", with the civilian clothes worn by insurgents

Weapons

Generally speaking, the Mexican Army was mostly equipped with British surplus weapons of the Napoleonic Wars; in fact, the standard infantry weapons were the Brown Bess musket for the line units and the Baker carbine for the light ones. Until becoming independent from Spain, Mexico had always produced large numbers of fine firearms for the colonial troops that were stationed on the territory of the Viceroyalty of New Spain. Soon after independence, however, the factories installed by the Spaniards in the country started to slow down their production and by 1834 the Mexican arsenals were only able to repair existing weapons but not to produce new ones. As a result of the above situation, the Mexican Army had no choice but to acquire all its new weapons from foreign sources. In 1825, ten years after the end of the Napoleonic Wars, Great Britain still had thousands and thousands of surplus weapons that had no chance of being employed by the British Army. The latter had been strongly reduced in its numbers after Waterloo and thus there were more flintlock muskets than soldiers. As a result, during the decades that followed 1815, Great Britain started to sell most of the oldest surplus weapons to the armies of different countries. The new Latin American republics, which had became independent from Spanish rule in recent times, were particularly interested in the purchase of British weapons; among them, Mexico was the single country that received more muskets and ammunitions from Great Britain. In 1825, thanks to some loans obtained from British banks, the Mexican government was able to acquire the following weapons from Great Britain: 90,000 Brown Bess muskets, 14,000 Baker carbines, 20,000 cavalry swords and 5,000 pistols for officers. With these weapons, the whole Mexican Army could be re-equipped with British arms. Generally speaking, these were in good conditions when they arrived in Mexico and they were still modern weapons by 1825. With the progression of time, however, flintlock muskets became soon outdated; by 1840 most of the European armies had replaced their old flintlock weapons with new percussion ones and thus the Brown Bess was completely outclassed as a line infantry musket.

Brown Bess Musket

In 1842 the US Army adopted its first percussion musket, the M1842 Springfield; this was still a smoothbore musket, but it was much more easy and rapid to reload if compared to the Mexican Brown Bess. During the years between the Texan Revolution and the US-Mexican War, the Mexican government did very little to update the arsenal of its troops and thus did not adopt new percussion weapons. In 1842 and 1844 another large stock of weapons was bought from Great Britain, but these were old surplus ones like those acquired in 1825: 20,000 Brown Bess muskets, 3,000 Baker carbines, 3,000 cavalry swords and 6,000 infantry short swords. The Brown Bess muskets, known as "Tower" ones since they bore the seal of the Tower of London where they had been produced, were of different models including the Pattern 1802 and the famous India Pattern. These weapons had all the same characteristics: walnut stocks, brass fittings, a weight between nine and twelve pounds, barrels with a length between 39 and 42 inches. The smoothbore barrel of the Brown Bess musket was of 0.753'' caliber but fired a ball of 0.71'' in diameter; the latter had a weight between one and three-sixteenths of an ounce. This large bullet had a great "stopping power", i.e. was always able to kill or wound quite seriously any man who was hit. The space located between the ball and barrel, known as "windage", helped to make the loading operations simpler and faster but greatly reduced the overall range/accuracy of the musket. The Brown Bess had no rear sights and thus was aimed simply by looking down the top of the barrel. The standard range of this weapon was between 100 and 700 yards, with a variation of up to 300 yards at any given elevation of the barrel. The effective range under optimum conditions was between 100 and 200 yards, with the deadliest effect only at 50 yards or less. As a result of the above performances, infantry tactics of the period did not encourage aimed fire by individuals but rather the output of a great volume of fire at a mass target. The rate of fire for an experienced fusilier could be as high as five rounds per minute and misfire was up to 15% of the time. Due to the tendency of the barrel to foul and thus require cleaning, it was generally very difficult to maintain the rate of fire described above. The standard bayonet of the Brown Bess had a length of about fifteen inches and consisted of a triangular-sectioned blade attached to a cylindrical socket that would fit on the end of the musket barrel. The socket-bayonets were held in place by a right-angled slot passing over the foresight of the musket and lacked any sort of locking ring. In addition, they did not have any form of handguard (differently from other contemporary models of bayonet).

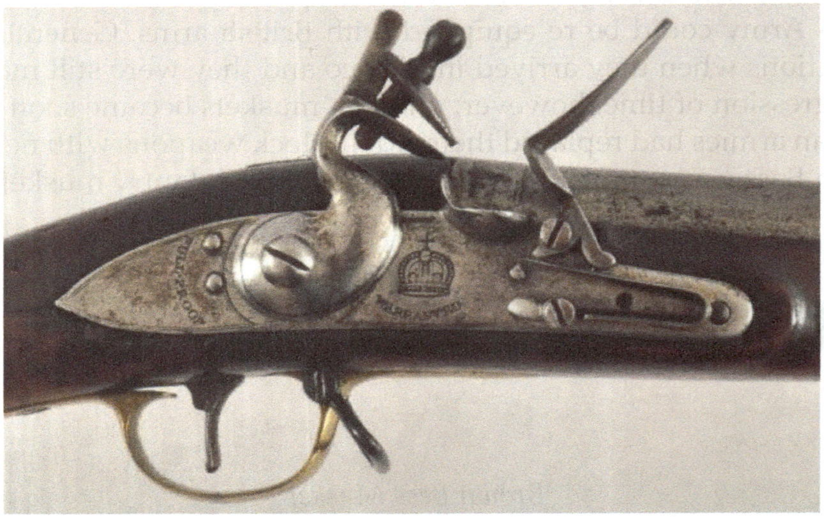

Detail of the Mexican Brown Bess flintlock mechanism

The "cazadores" companies of the line infantry regiments and the four light infantry regiments were all armed with the Baker carbine; this was an excellent precision weapon, with a total length of 46 inches and a short barrel of 30 inches. The latter had a weight of nine and one-half pounds. The caliber of the carbine could vary from 0.615'' to 0.70'' and the barrel had a seven-groove rifling that made one turn in 120 inches. In order to reload his weapon, the single rifleman had to place the butt of the carbine between his feet to ram the bullet down the barrel by using both hands. The accuracy of the Baker carbine was much better than that of the Brown Bess musket, since it had an effective range of up to 200 yards. The rate of fire and the general reliability of the carbines, however, were inferior to those of the muskets due to a lower tolerance for fouling in the barrel. The Baker carbine had a sword-bayonet with broad blade, having an overall length of 23 inches and a brass handle that was attached to the side of the barrel by spring-clips.

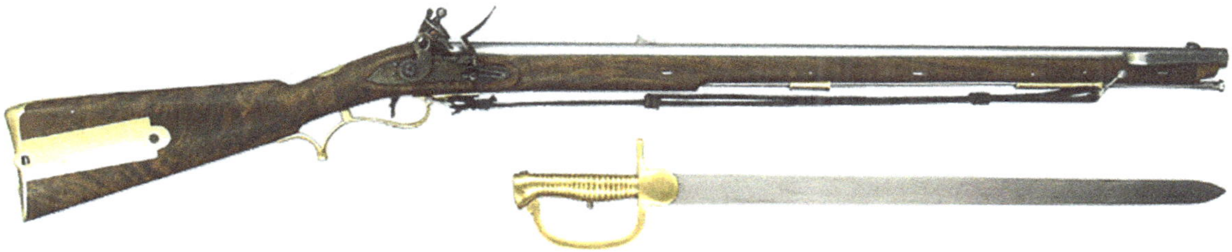

Baker carbine with related sword-bayonet

Mexican cavalry units were armed with sabers, carbines, pistols and lances. All these weapons, except the lances, were imported in great numbers from Great Britain. Judging from contemporary sources, however, large quantities of surplus sabers dating back to the Napoleonic Wars were imported also from France. Usually the cavalry carbines were a shortened version of the Brown Bess muskets, but sometimes cavalry units could be equipped with rifled Baker carbines. In addition to the above, some locally-produced or locally-modified "escopetas" were also in use. These were short and smooth-bore weapons, looking bell-mouthed. They fired a heavy bullet that was extremely effective if shot from a short distance. The accuracy of the "escopetas" was quite scarce, especially because they were frequently fired from above the head in order to avoid the tremendous recoil of the weapon. The muzzle of the cavalry carbines had a deep funnel to allow the ball and patch to be held while the ramrod was being drawn for loading on horse. Usually each Mexican cavalryman also had one or two flintlock pistols, which were employed only in case of need and from a very short distance.

Brown Bess cavalry carbine

Irregular and semi-regular cavalry were also armed with deadly "lazos", a very peculiar weapon that was quite effective if used by skilled hands. The favourite weapon of the Mexican mounted troops was with no doubts the lance; this consisted of a shaft made of beech or nut wood plus a point with socket made of metal. The point was eight and one-quarter inches long, having three or four cutting edges with concave gutters. At the lower end of the point there was a metal crosspiece followed by a tube and two iron straps that were screwed onto the shaft as a protection against saber cuts that could chop off the point. The lance also bore a pennon, which had a double-pointed shape; this was usually in the distinctive colours of each cavalry regiment. The pennon did not have only a "symbolic" function, since it scared the enemy's horses by fluttering near their eyes. A leather sling was attached to the shaft of the lance to allow each cavalryman to suspend the weapon from the right arm or to support a thrust. Usually the right stirrup of lancers had a leather tube for resting the lance and the cinch had a loop that could be used to make sure that the lance was not lost while riding the horse. According to contemporary sources, the American cavalrymen who fought in close-combat with Mexican lancers usually had very little hopes of success; the irregular "vaqueros" of the Mexican Army, in fact, employed their lances in a very effective way and were skilled light cavalry skirmishers. Also the regular cavalrymen, however, employed their lances on several occasions in order to break the enemy formations with a frontal charge.

12-pdr cannon mounted on a Gribeauval carriage

The Mexican artillery was almost entirely formed by ageing brass Spanish guns dating back to the colonial period, since the government of Mexico City was never able to buy modern pieces from the European nations as it did for the light weapons of the infantry and cavalry. In addition, there were no industries on the Mexican territory that could produce modern guns in an autonomous way. As a result of the above, by the outbreak of the war with the United States, the Mexican artillery could deploy only 150 field guns in addition to the dozens of heavy pieces that were scattered across the fortresses of the national territory. These field guns were all brass pieces of 4-pdr, 6-pdr, 8-pdr and 12-pdr mounted on Gribeauval carriages. Spain had been one of the first nations in the world to adopt the new Gribeauval artillery system and thus the guns employed by the Mexican Army were of good quality, but they were becoming rather outdated if confronted with those deployed by the much more updated US artillery. In practice, the Mexican Army went to war with weapons that would have been perfect for the Napoleonic battlefields of 30 years before but that were now in the process of being outclassed by more modern technologies. The main arsenal of the Mexican artillery, located at Mexico City, could repair the existing pieces but was not able to produce new ones; as a result, the Mexicans had no chance of replacing the guns that were destroyed on the battlefield or that were captured by the enemy. In 1846, due to the state of emergency determined by the war, there were some attempts to cast brass field guns locally; the few pieces that were produced, however, soon proved to be of disappointing quality. If compared to the American one, the Mexican artillery had many weak points: first of all, Mexican black powder was of scarce quality and this led to the production of less effective ammunitions; secondly, the Mexican field pieces were moved only with great difficulty and thus could not be deployed where needed in a rapid way. The horses of the Mexican artillery train were too small and light to transport heavy pieces; in addition, the movements of guns were done by private contractors and not by military personnel. These civilians had no combat experience and no discipline to speak of; on most occasions they did not employ horses but oxen or mules and this slowed down a lot the movements of the artillery train. During a battle, the Mexican guns usually occupied always the same position since it was practically impossible to move them under enemy fire. The heavy pieces mounted in fortresses, usually brass guns of 16-pdr or 24-pdr, were not particularly effective since most of them were too old. However, there were some exceptions to this rule: at Vera Cruz, for example, the Mexicans had 70 British iron guns including several efficient 24-pdrs and 32-pdrs. On several occasions, the Mexican artillery also employed Congreve rockets purchased from Great Britain or local copies of them.

US uniforms and plate commentaries

The US Army of 1845 wore the uniforms prescribed by the dress regulations that had been promulgated on 21 June 1839; these had substituted the previous ones issued during 1836 and were slightly modified in 1841. During the war, the dress regulations were modified on two occasions: in 1846, when the new uniforms for the "Regiment of Mounted Riflemen" and for the "Company of Sappers, Miners and Pontoniers" were added; in 1847, when the new uniforms for the "Regiment of Voltigeurs and Foot Riflemen" were added. The 1839 dress regulations, which remained valid until 1851, were extremely complete; in fact, they prescribed three main different kinds of uniform: parade dress, winter service dress and summer service dress. The latter consisted of an entirely white summer uniform that was designed for use during the hot months; this could have been extremely practical for use in Mexico, but by the time hostilities broke out it had been distributed only to a limited number of units. As a result, it was the winter service dress that was mostly employed by the US troops in Mexico. The uniforms of the numerous volunteer units were not provided by the central government, since it was the responsibility of each state to produce uniforms for its volunteer corps. Generally speaking most of the volunteer units wore uniforms that were exact copies of those prescribed by the official dress regulations for the regular army; some volunteer corps, however, had distinctive uniforms that were quite different from those of the regulars. A good number of the volunteers did not have a proper uniform at all; in addition, with the progression of the war, it was common practice for both the regulars and the volunteers to mix what remained of their original uniforms with Mexican civilian clothes (including "ponchos" and "sombreros").

Officers: The headgear prescribed for parade dress by the uniform regulations of 1839 was a black cocked hat with golden loop and tassels for senior officers and a black shako with brass frontal plate for junior officers (like the one reproduced in figure C3). Sometimes the cocked hat also had a decorative plume, in different colours according to rank. In practice, however, for the entire duration of the war against Mexico all US officers (like their men) wore the comfortable and practical M1839 soft "forage cap" represented in figure A1. This was entirely dark blue, with black leather visor and chinstrap. A button on each side of the cap held the leather chinstrap in place. Senior officers wore a dark blue double-breasted frock coat, having two rows of eight gilt buttons each on the front. This had turnbacks only in the parade version. The standing collar and round cuffs of the coat were entirely dark blue. Rank was shown by golden "epaulettes" on the parade frock coat and by golden shoulder boards on the service one. Junior officers, instead, had a single-breasted coat with one row of eight to ten buttons on the front. The latters were made of white metal, while both the "epaulettes" and the shoulder boards of the junior officers were silver ones. Senior officers wore entirely dark blue trousers during cold months and entirely white ones during hot months; junior officers, instead, had light blue trousers with a single stripe in their branch of service's colour during winter months. Infantry had white as distinctive colour, cavalry had yellow and artillery had red. Senior officers wore a yellow waist-sash, while junior ones had a red waist-sash. The uniforms described above were obviously employed also by the officers of the Engineer Corps. A good example of senior officer's service dress is reproduced in figure C1. All the officers of the Marine Corps, be they senior or junior ones, had double-breasted frock coats.

A3
A1
A2
Benedetto Esposito 2019

Infantry: As parade headgear for the infantry, the 1839 dress regulations prescribed a black shako with plume in the distinctive branch of service's colour of each unit. On the front, this had two metal badges: one, made of brass, reproduced the American Eagle; the other, made of white metal, reproduced the distinctive symbol of each branch of service. As a result of the above, an infantry shako had white plume and a white metal badge reproducing a bugle horn. During the whole war with Mexico, however, the shako was rarely seen and was always replaced by the forage cap. The latter was to have a bottom band in the distinctive colour of each branch of service, but apparently all infantrymen of the US Army usually wore an entirely dark blue one (see figure A1). The parade uniform, never used during the war, consisted of a dark blue single-breasted frock coat with white piping to front and cuff flaps. This had double white frogging on the collar and cuff flaps; in addition, it had white "contre-epaulettes" and white piping to the turnbacks. Trousers were light blue, with white stripe. The winter service dress, which was widely worn during the conflict, looked exactly like the one represented in figure A1. It was entirely light blue and had simple shoulder straps piped in white instead of the "contre-epaulettes". Collar had the same white frogging of the parade dress, but the round cuffs were simple and did not have patches. The light blue trousers had a white stripe. Both the jacket and the trousers were made of kersey cloth. Sergeants wore the same red waist-sash of junior officers. Rank was shown by white upward pointing "chevrons", reproduced on both sleeves; years of service were shown by dark blue upward pointing "chevrons" applied near the cuffs. An entirely white summer version of the uniform described above did exist and would have been perfect for the climate of Mexico, but apparently this was issued to just a few units of the regular army. The two crossbelts and the waistbelt worn by each infantryman were made of white leather and had brass buckles bearing the inscription "US". Each soldier also had a cotton haversack and a canteen. The latter was suspended to a brown leather crossbelt and was made of sheet iron painted in light blue or grey. The elite "Regiment of Voltigeurs and Foot Riflemen", at least according to official dress regulations, was to wear an entirely light grey uniform which had all the same features of the line infantry one. Unfortunately, however, these uniforms were all lost at sea while being transported to Mexico; as a result, the regiment had to adopt the entirely dark blue dress reproduced in figure A3.

B3
B1
B2
Benedetto Esposito 2019

Cavalry: During the war against Mexico the US Dragoons always wore their campaign uniform. This, reproduced in figure B1, included a dark blue forage cap with yellow bottom band. The dark blue single-breasted jacket had yellow piping to pointed cuffs, shoulder straps, standing collar, front and bottom edges. Also the rear seams of the jacket were piped in yellow. In addition, it had yellow frogging on the collar. Rank was shown in the same way as for the infantry, but with yellow "chevrons" instead of white ones. Trousers were light blue with yellow stripe. Brass company letters could be sometimes worn on the front of the forage cap. All belts were made of white leather while the ammunition pouch, as for the infantry, was made of black leather. The elite "Regiment of Mounted Riflemen" wore the peculiar uniform represented in figure B3; this was very similar to that of the dragoons, but had dark blue trousers instead of light blue ones. The famous "Spy Company" formed with Mexican renegades, instead, wore the dress reproduced in figure B2: this included a green shell jacket with red pointed cuffs (probably obtained from captured Mexican cavalry uniforms) and a red band wrapped around the hat. All the rest of the "uniform" was made up by pieces of Mexican civilian dress. Most of this unit's members were armed with lances and these had entirely red pennons.

C3
C1
C2
USM
USM
USM
US
Benedetto Esposito '19

Artillery: The winter service dress of the US artillery is reproduced in figure C2. This consisted of a dark blue soft cap with red bottom band, light blue shell jacket with yellow piping/frogging to standing collar and light blue shoulder straps piped in yellow, light blue trousers with red stripe. As a result of the above, it looked like a "mix" between the infantry and cavalry uniform. Mounted batteries were dressed exactly like the foot ones, except for Major Ringgold's own Company C of the 3rd US Artillery which had a distinctive dress. This included a black shako with red falling plume, having two brass badges on the front: one represented an American Eagle and the other reproduced two crossed cannons. In addition, the shako also had red cords wrapped around it. The shell jacket was dark blue instead of light blue and it had red piping/frogging to standing collar; also the dark blue shoulder straps were piped in red and not in yellow. Trousers were the same of the other artillery companies. Personal equipment was exactly the same used also by the line infantry.

Sappers, Miners and Pontoniers: This single company wore the elegant parade dress reproduced in figure C3. This included a black shako with black pompom and two brass badges on the front; one reproduced an American Eagle, while the other represented a castle (the distinctive symbol of the unit). The dark blue single-breasted frock coat had a single line of yellow frogging on the collar and dark blue turnbacks. The number of yellow frogging lines on the cuffs corresponded to rank; the latter was also shown by yellow inverted "chevrons" on the sleeves. In our case the three frogging lines on the cuffs and the three inverted "chevrons" on the sleeves are typical of a sergeant. Privates wore yellow "contre-epaulettes" on their shoulders, while NCOs had "epaulettes" of the same colour. On campaign, the "Company of Sappers, Miners and Pontoniers" received the same service dress of the artillery but with dark blue jacket and trousers.

Marine Corps: The service dress of Marine officers was that reproduced in figure C1. Enlisted men wore the same uniform of the line infantry (figure A1), but without any of the white lace that adorned infantry shell jackets. Also the light blue trousers did not have the single white stripe. Rank was shown by yellow stripes and not by white inverted "chevrons".

D3
D1
D2
US
US
US
Benedetto Esposito 2019

Illinois Volunteers: According to the will of their Governor, the volunteers coming from Illinois should have been dressed with blue shell jackets and trousers made of jeans cloth; apparently, however, only some units received this dress and soon abandoned it after arriving in Mexico. The volunteers, in fact, did not like the dress of the regulars or simpler versions of it. During the campaign in Mexico each company of Illinois volunteers had its own distinctive uniform, which generally comprised a short jacket that could be produced in various shades of blue or grey. This usually had red or yellow facings and could be adorned by some frontal frogging in these same colours or in white. This kind of uniform employed by the volunteers from Illinois is represented in figure D3. Apparently several volunteer units, wishing to "enrich" their uniforms that were copies of those worn by the regulars, applied some coloured frogging on the front of their shell jackets.

Indiana Volunteers: The governor of Indiana did not prescribe any specific uniform for the volunteer units raised in his state, but ordered that all volunteer officers were to have the same dress of their regular equivalents. Usually the volunteers from Indiana wore double-breasted shell jackets and trousers made of grey or blue cloth, together with standard forage caps.

Kentucky Volunteers: The official dress consisted of dark blue shell jacket and trousers in the style of the regulars' uniforms, but this was not employed by all units since there was a chronic shortage of dark blue cloth.

Massachusetts Volunteers: The single unit coming from this state wore a very simple but practical uniform consisting of a single-breasted grey coat with white collar and cuffs.

Mississippi Volunteers: The famous regiment commanded by Jefferson Davis wore the distinctive uniform reproduced in figure D1. This was very similar to that employed in 1836 by the "Alabama Red Rovers", a unit of volunteers coming from Alabama that fought against the Mexicans during the Texas Revolution. It consisted of a red double-breasted blouse, usually worn with white trousers and a slouch hat made of straw. Apparently all the volunteers under command of Davis were equipped with a deadly "Bowie knife", which they employed with great success on several occasions. Differently from the majority of the volunteers, who were armed with old flintlock muskets, those under command of Davis were equipped with percussion rifles.

Missouri Volunteers: Most of the volunteers coming from this state wore no uniforms at all; a notable exception to this rule was represented by one of the artillery companies, which was dressed quite similarly to the regular artillerymen. The uniform comprised a dark blue forage cap with red bottom band, dark blue jacket with red collar and dark blue trousers with single red stripe.

Mormon Battalion: The Mormon volunteers went to California dressed with their own civilian clothes, but a good number of them gradually started to wear a simple "uniform" consisting of dark blue forage cap and grey single-breasted jacket. Brown trousers were quite popular. This simple dress is reproduced in figure D2.

New York Volunteers: The original uniform of the 1st Regiment of New York Volunteers, represented in figure A2, was quite elegant. It comprised a dark blue shako with yellow piping and black leather visor, dark blue single-breasted jacket with red pointed cuffs and collar patches, grey trousers with single red stripe. When the regiment reached California, new shell jackets with red collar and pointed cuffs were worn together with white trousers. This new uniform, much simpler than the previous one, included a dark blue forage cap instead of the shako.

Ohio Volunteers: Basically the Ohio volunteers wore the same standard uniform of the regulars, including a single-breasted dark blue jacket with loops of worsted tape trim on the cuffs.

Pennsylvania Volunteers: The volunteers from this state wore the same uniforms of the US regular infantry, but these were made with dark blue cloth instead of light blue one.

South Carolina Volunteers: The so-called "Palmetto Regiment" wore the same basic dress of the regulars, made with dark blue or grey cloth. This had two peculiar features: wing-style shoulder straps and pointed white cuff lace. The prescribed uniforms of the "Palmetto Regiment", however, were lost while being transported to Mexico and thus the South Carolina volunteers were given light blue regular uniforms comprising visored caps with a frontal badge reproducing a palmetto tree.

Texas Volunteers: Most of the volunteers coming from Texas, including the famous Rangers, wore their everyday civilian clothes during the whole conflict. These comprised buckskin "hunting shirts" and high boots. Most of the "Diablos Tejanos" were armed with Colt revolvers and Bowie knives, in addition to short carbines that were employed to fire from short distance. With their bearded faces, uncut clothes and harsh manners they were not so different from their Mexican opponents (the irregular "guerrilleros"). A nice example of the standard outfit worn by Texas Rangers is reproduced in figure J3.

Virginia Volunteers: The original uniform prescribed for the volunteers coming from Virginia consisted of forage cap, shell jacket and trousers all obtained from dark blue cloth. After six months of service, however, most of the uniforms had been destroyed by prolonged use and thus were replaced by new ones that were exact copies of those worn by the regulars.

Mexican Uniforms and Plate Commentaries

Guard Units: The Foot Grenadiers of Iturbide wore a distinctive red uniform with green facings. This consisted of: black bearskin cap with brass back plate bearing a Mexican Eagle, red cords and flounders; red frock coat with green lapels, collar, round cuffs and shoulder wings; yellow piping to collar and "sardinetas" on the sleeves (i.e. three pointed-end lines of yellow lace), shoulder straps made of brass scales, white trousers. The Horse Grenadiers of Iturbide, instead, had the following hussar-style uniform: black colpack with white metal chinscale, plume and side-bag in the Mexican national colours; white "dolman" with red collar and pointed cuffs piped in green, green frontal frogging and three rows of white metal buttons; green "pellisse" with external edging of black fur and white frogging with white metal buttons; half-white and half-green waist-sash; green trousers with white "Hungarian knots" on the front and single stripe. The Foot Grenadiers of Santa Anna wore the elegant uniform with black bearskin cap and red frock coat reproduced in figure E1; the Hussars, instead, wore the ornate dress with black colpack and red "dolman" represented in figure E2.

Infantry: The first official dress regulations of the Mexican Army, promulgated on 20 September 1821, prescribed a very simple "Napoleonic-style" uniform for the infantry. This was to consist of the following elements: black shako with pompom and cockade in the Mexican national colours, brass frontal plate bearing the unit number over the Mexican coat-of-arms and chinscale; dark blue single-breasted frock coat with red collar, round cuffs, cuff flaps, turnbacks, piping to front and to shoulder straps; grey or dark blue trousers (white during summer). Sometimes a white cord with flounders was wrapped around the shako, which could also have yellow top and bottom bands. The red collar of the coat could have a brass unit number; the latter was reproduced also on the brass buttons of the coat. This same dress was also worn by the infantrymen of the Active Militia. Sappers wore massive bearskin caps and leather aprons. Grenadiers and "cazadores" were easy to distinguish thanks to the presence of two distinctive elements on their uniforms: "epaulettes" on the shoulders (red for grenadiers and green for "cazadores") and "sardinetas" on the sleeves. In addition, the pompom of the shako was red for grenadiers and green for "cazadores". On campaign, most soldiers replaced their shakos with dark blue fatigue caps ("bonnets de police") having red piping and frontal tassel. During summer, most of the infantrymen wore their usual frock coat but in white colour instead of dark blue. For a brief period, during January 1832-June 1833, some alterations were made to the general uniform described above: red lapels were added to the front of the coat and white piping was added to collar, cuffs, turnbacks and lapels. After a short experimentation, however, these modifications were abandoned and the uniforms remained unchanged until 1839. As a result, it was with the simple dress described above that the Mexican infantry took part to the Texas Revolution.

The new dress regulations of 10 July 1839 prescribed radically new uniforms based on regimental distinctions, but these were not adopted in a definitive way until 22 December 1841. The new uniforms were quite similar to those adopted for a short period during the "experimentation" of 1832/1833, since they included coloured lapels. Coats and trousers were dark blue for all units, but each regiment had a distinctive combination of colours for the facings (collar, cuffs and cuff flaps), lapels and piping. 1st Regiment: red collar, cuffs and cuff flaps; yellow lapels and piping; 2nd Regiment: light blue collar; red lapels, cuffs, cuff flaps and piping; 3rd Regiment: light blue collar and piping; red lapels, cuffs and cuff flaps; 4th Regiment: light blue collar and cuffs; white piping; red lapels and cuff flaps; 5th Regiment: light blue cuffs and piping; red collar, lapels and cuff flaps; 6th Regiment: red collar, cuffs and cuff flaps; white lapels; light blue piping; 7th Regiment: green collar and cuffs; red lapels, cuff flaps and piping; 8th Regiment: red collar and cuffs, light blue lapels and cuff flaps, white piping; 9th Regiment: buff collar and cuff flaps, purple lapels and cuffs, white piping; 10th Regiment: red collar and cuffs, purple lapels and cuff flaps, buff piping; 11th Regiment: red collar, cuffs and cuff flaps; green lapels, white piping; 12th Regiment: buff collar, lapels and cuffs; red cuff flaps, white piping. The general appearance of the new uniforms introduced in 1841 was that reproduced in figure F1. Shortly before the outbreak of the war with the United States, most of the dark blue frock coats of the Mexican infantry started to be replaced by new tunics ("levitas") of the same colour. The tunic, which was not prescribed by any official dress regulations but that soon became quite popular, was much more comfortable than the frock coat: instead of a front rounded off at the waist and knee-length tails turned back behind the legs, in fact, it had a full skirt that reached the knees all around and was single-breasted (it had no lapels). Together with the "levita", a certain number of Mexican infantrymen also started to wear a new model of shako: this, differently from the previous cylindrical one, was tapered and had a rear profile sloping forward. The uniforms of the Active Militia continued to be the same of the regular infantry until April 1842, when a distinctive dress was specified also for them. This was as follows: black shako with red pompom; dark blue coat with red collar, cuffs, lapels and turn-backs; brass buttons, yellow piping, light blue trousers with single red stripe (see figure J2). This same uniform was adopted also by the infantry of the Coast Guard Militia, which had used the same dress of the regulars until 1842. The Battalion of Invalids wore the following uniform: black shako; dark blue frock coat with light blue collar, cuffs, lapels and turn-backs; red piping. On the collar, in red, there was the inscription "INVALIDOS". The new light infantry regiments used two different uniforms: the original three raised during 1840-1843 received a very "modern" grey dress, including peaked cap and "levita" tunic (represented in figure F3); the fourth regiment, instead, used a different dress with "old-style" frock coat that was practically identical to the standard line infantry uniform (reproduced in figure F1). This included "sardinetas" on the cuffs and "epaulettes" on the shoulders for elite companies. All light infantry regiments had black belt equipment, differently from the line ones which had white belt equipment. The Battalion "Fijo de Mexico" wore the distinctive uniform reproduced in figure F2; this was quite similar to the standard dress worn by all line infantry units during summer, but had light blue trousers instead of white ones. The "Fixed Battalion of the Californias" wore an entirely dark blue uniform, represented in figure I2. Officers, like in all the Mexican military units, wore brass "gorgets" on the chest and had golden "epaulettes"/"contre-epaulettes" on the shoulders showing their rank.

F3
F1
F2
Benedetto Esposito '19

Cavalry: According to the dress regulations promulgated on 20 September 1821, the Mexican line cavalry was to wear the following uniform: brass helmet with white metal frontal plate reproducing a Mexican Eagle over the regimental number, white metal chinscale, black "caterpillar" comb, white metal crest and furniture; yellow single-breasted frock coat with red collar, pointed cuffs, turnbacks and frontal piping; white metal buttons and unit number on the collar, dark grey trousers with single red stripe. On 6 December 1824 this dress, which had retained the distinctive yellow colour of Spanish cavalry, was replaced by a new one having red as main colour: black leather helmet with white metal frontal plate reproducing a Mexican Eagle over the regimental number, white metal chinscale, black mane with a pompom in the Mexican national colours at the front, plume in the Mexican national colours, white metal crest and furniture; red single-breasted frock coat with green collar, pointed cuffs, turnbacks and frontal piping; white metal buttons and unit number on the collar, dark grey or dark blue trousers with single red stripe. With both uniforms officers showed their rank by wearing silver "epaulettes"/"contre-epaulettes" on the shoulders. According to the new dress regulations of 1824, the cavalry units of the Active Militia were to wear the same uniforms of the regulars but with inverted colours: as a result, their green single-breasted frock coat had red collar, pointed cuffs, turnbacks and frontal piping. The helmet was of the old model with black "caterpillar" comb (albeit being made of black leather instead of brass). Like their infantry equivalents, the "zapadores" of mounted units wore black bearskin caps. On 2 January 1832 the red uniforms were slightly modified with the addition of green frontal lapels.

With the new dress regulations of 10 July 1839, each cavalry regiment received a distinctive uniform like the infantry ones; the new dress, however, started to be worn only since 22 December 1841. During this short period of transition, all cavalry regiments were given a simple "unified" uniform comprising the following elements: black shako with white top and bottom band, white metal frontal plate bearing Mexico's coat-of-arms and the regimental number, red pompom and cockade in the Mexican national colours; light blue single-breasted frock coat with red frontal piping, collar, pointed cuffs, turnbacks and "contre-epaulettes"; dark blue trousers strapped with leather and having a single red stripe. This uniform is reproduced in figure G3. The new "regimental" dress issued in late 1841 had some elements that were common to all units: black shako with white top and bottom band, white metal frontal plate bearing Mexico's coat-of-arms and the regimental number, red pompom and cockade in the Mexican national colours; dark blue trousers strapped with leather and having a single red stripe. Each regiment, instead, had its own distinctive coat; this new uniform it represented in figure G1. 1st Regiment: yellow coat with red collar, cuffs, lapels, "contre-epaulettes" and turnbacks; 2nd Regiment: yellow coat with blue collar, cuffs, lapels, "contre-epaulettes" and turnbacks; 3rd Regiment: blue coat, green collar and cuffs; white lapels, "contre-epaulettes" and turnbacks; 4th Regiment: light blue coat with red collar, cuffs, lapels, "contre-epaulettes" and turnbacks; 5th Regiment: dark blue coat with red collar, cuffs, lapels, "contre-epaulettes" and turnbacks; 6th Regiment: green coat; white collar, cuffs, lapels, "contre-epaulettes" and turnbacks; 7th Regiment: white coat; light blue collar, cuffs, lapels, "contre-epaulettes" and turnbacks; 8th Regiment: dark blue coat; red cuffs, lapels, "contre-epaulettes" and turnbacks; white collar; 9th Regiment: green coat; red collar, cuffs, lapels, "contre-epaulettes" and turnbacks. On 7 September 1845, the 1st Regiment received the new peculiar uniform reconstructed in figure G2; this comprised a black leather "Tarleton" helmet with brass frontal plate, black crest and red plume. The green coat had

G3
G1
G2
Benedetto
Esposito '19

yellow piping to collar and cuffs; yellow cuff flaps, lapels and turnbacks; green "epaulettes" piped in yellow and having red fringes. Trousers were grey, strapped with leather and having a single red stripe. Around 1845 most of the Mexican cavalry regiments had already abandoned their "regimental" uniforms described above and were wearing a new "unified" dress (which was not prescribed by official regulations). This was as follows: black shako with yellow top and bottom band, brass frontal plate bearing Mexico's coat-of-arms and the regimental number, pompom and cockade in the Mexican national colours; dark blue single-breasted tunic with red frontal piping, collar, round cuffs, short turnbacks and shoulder rolls; light blue trousers with single red stripe.

Until 31 August 1840 all the Mexican light cavalry units wore the following uniform: black leather helmet with brass frontal plate and crest; dark blue single-breasted frock coat with red piping to collar, cuffs, turnbacks and front; grey trousers with single red stripe. In 1840, however, this was abandoned and the same uniform employed by the line cavalry was adopted; the latter, reproduced in figure G3, was retained also when the line cavalry received new "regimental" dress. The ornate uniform of the "Tulacingo Cuirassiers", including brass helmet and cuirass, in represented in figure H3; the red one of the "Jalisco Lancers", including a Polish "czapska" headgear, is reconstructed in figure H2. The "Vera Cruz Lancers" wore this same dress, but with inverted colours (green with red facings). The green uniform of the "Mounted Chasseurs", with black colpack and red frontal plastron decorated with white frogging, is reproduced in figure H1. Regarding the mounted units of the Active Militia and Coast Guard Militia, until 1840 they continued to wear the green uniform with red facings described above; during that year, however, they received the same light blue dress given to the light cavalry units (reproduced in figure G3). The Presidial Cavalry Companies continued to wear their distinctive and simple uniform also after Mexico became independent from Spain; this is represented in figure I1. It comprised a black wide-brimmed round hat with a white band; dark blue single-breasted jacket with red collar, cuffs and short turnbacks; dark blue or grey trousers with single red stripe. While serving against the natives, the "Cuera soldiers" wore protective leather coats over their uniforms and sometimes also employed small leather shields. In January 1842 the letters "AC" (Alta California) and "BC" (Baja California) were added to the collars of the Presidial Companies stationed in Upper California and Lower California. The irregular "Auxiliary Cavalry Companies" and "guerrilleros" of the Mexican Army all wore their everyday civilian clothes of "rancheros" (cowboys), which included "sombreros" and "ponchos" as well as white lace "calzoncillos" that were worn under leg-protectors made of leather. The complete outfit of a Mexican irregular lancer is reconstructed in figure J1.

Artillery and technical corps: Until 1840 the Mexican artillery wore a very simple uniform that was quite similar to that of the regular infantry. This comprised the following elements: black shako with yellow top and bottom band, brass frontal plate and chinscale, red cord and flounders, national cockade and red plume; dark blue single-breasted frock coat with red collar, round cuffs, turnbacks, "epaulettes", piping to front and to dark blue cuff flaps. Brass flaming grenade on collar. Dark blue trousers with single red stripe. On 31 August 1840 a new uniform was introduced, which was as follows: black shako with yellow top and bottom band, brass frontal plate (showing a Mexican Eagle over two crossed cannons) and chinscale, national cockade and red pompom; dark blue frock coat with red collar, round cuffs, turnbacks and piping to dark blue cuff flaps. The coat had black lapels with yellow-laced buttonholes and yellow "epaulettes". Yellow flaming grenade on collar, bearing unit number. Dark blue trousers with single red stripe. Horse artillery wore this same uniform but with shorter tails and trousers strapped with black leather. Until 1840 the "Zapadores" wore the same uniforms of the artillery, albeit with two crossed axes on the collar instead of the flaming grenade. In that year they received a new dress, which was as follows: black shako with yellow top and bottom band, brass frontal plate (showing a Mexican Eagle over two crossed axes) and chinscale, national cockade and red pompom; dark blue frock coat with black collar and lapels, red round cuffs, turnbacks and piping. Yellow crossed axes on collar. Dark blue trousers with single red stripe. Until 19 November 1842 the Mexican Marine Corps had the same uniform of the line infantry for its companies of naval infantry and the same uniform of the foot artillery for its company of naval artillery; only the yellow unit badge embroidered on collar was different, consisting of a simple anchor for naval infantry and of an anchor across a flaming grenade for naval artillery. In 1842 Santa Anna gave to his Marine Corps the flamboyant green uniform reconstructed in figure E3 (with the "sardinetas", "epaulettes" and additional lace typical of elite troops). The cadets of the "Colegio Militar" had a parade dress consisting of black shako, dark blue single-breasted frock coat with red piping to front and turnbacks, red collar and round cuffs piped with gold lace, gilt buttons and white trousers. In 1840 the colour of collar and round cuffs was changed to light blue. For winter service, the cadets wore a dark blue "bonnet de police" fatigue cap having red piping and yellow lace together with a grey greatcoat having red collar patches. Sometimes also the cuff flaps of the greatcoat could be red; this winter uniform is reproduced in figure I3 and comprised grey trousers with single red stripe.

Civic Militia, National Guard and foreign volunteers: Civic Militia uniforms were not compulsory and thus only a limited number of the militiamen usually wore them. According to the dress regulations of 1822, the Civic Militia infantry units were to use the following uniform: black shako; light blue single-breasted frock coat with yellow collar, cuffs and piping; light blue trousers with single red stripe. Cavalry had this same uniform, while artillery had red collar and cuffs plus an additional yellow flaming grenade on collar. In 1823 the colour of the coat was changed to dark blue for all units, while in 1833 the colour of the coat for cavalry units was changed to red. The units from the Federal District of Mexico City bore the inscription "DISTRITO FEDERAL" on the collar. The "Commerce Battalion of the Federal City" had the following peculiar uniform: black shako with yellow top and bottom band, brass frontal plate and chinscale, red pompom and cockade in the national colours of Mexico; dark blue single-breasted frock coat with black collar and round cuffs piped in red, yellow "sardinetas" on the sleeves, red "contre-epaulettes" on the shoulders, black turnbacks piped in red, yellow frogging and red piping on the front. Dark blue trousers. The National Guard battalions of the Federal District, which fought heroically during the defence of the Mexican capital, had a very "modern" dress that comprised the following elements: black shako with yellow bottom band, brass frontal badge bearing the unit's name, brass chinscale, red pompom and national cockade; dark blue tunic ("levita") with red collar, round cuffs, cuff flaps, "epaulettes" and piping to front; brass buttons and flaming grenade on collar; red trousers and white spats. The National Guard units from other areas of Mexico retained their previous Civic Militia uniforms or wore their civilian clothes. The famous "Saint Patrick's Battalion" wore the same dress prescribed for the foot units of the Active Militia in April 1842: black shako with red pompom; dark blue coat with red collar, cuffs, lapels and turn-backs; brass buttons, yellow piping, light blue trousers with single red stripe. This uniform is represented in figure J2.

I3
I1
I2
AC AC
F.C F.C
Benedetto Esposito '19

J3
J1
J2
- Benedetto Esposito 2019 -

Images of the Major Battles of the US-Mexican War

Battle of Palo Alto, 8 May 1846

Battle of Monterrey, 21-24 September 1846

Battle of Buena Vista, 22-23 February 1847

Siege of Vera Cruz, 9-29 March 1847

Battle of Cerro Gordo, 17-18 April 1847

Battle of Contreras, 20 August 1847

Battle of Churubusco, 20 August 1847

Battle of Molino del Rey, 8 September 1847

Battle of Molino del Rey, 8 September 1847

The Storming of Chapultepec Castle, 12-13 September 1847

The Storming of Chapultepec Castle, 12-13 September 1847

The US Army entering Mexico City, 14 September 1847

Select Bibliography

Anton Adams, The War in Mexico, Emperor's Press, 1998

René Chartrand, Santa Anna's Mexican Army 1821-1848, Osprey Publishing, 2004

Ron Field, US Army Frontier Scouts 1840-1921, Osprey Publishing, 2003

Philip Haythornthwaite, The Alamo and the War of Texan Independence 1835-1836, Osprey Publishing, 1986

Stephen L. Hardin, The Texas Rangers, Osprey Publishing, 1991

Stephen L. Hardin, The Alamo, Osprey Publishing, 2001

Joseph Hefter, El Soldado Mexicano 1837-1847, Ediciones Nieto-Brown-Hefter, 1958

Joseph Hefter, Cronica del Traje Militar en Mexico del Siglo XVI al XX, Artes de Mexico, 1968

Joseph Hefter, The Army of the Republic of Texas, Old Army Press, 1971

Joseph Hefter, The Navy of the Republic of Texas, Old Army Press, 1974

Alan C. Huffines, The Texas War of Independence 1835-1836, Osprey Publishing, 2005

Hunt Janin and Ursula Carlson, The California Campaigns of the US-Mexican War 1846-1848, McFarland & Company, 2015

Philip Katcher, The Mexican-American War 1846-1848, Osprey Publishing, 1989

Philip Katcher, US Infantry Equipments 1775-1910, Osprey Publishing, 1989

John Langellier, US Dragoons 1833-1855, Osprey Publishing, 1995

Bruce Marshall, Uniforms of the Republic of Texas 1836-1846, Schiffer Publishing, 1999

Bruce Marshall, Uniforms of the Alamo and the Texas Revolution, Schiffer Publishing, 2003

Douglas V. Meed, The Mexican War 1846-1848, Osprey Publishing, 2002

Clayton R. Newell, The Regular Army before the Civil War 1845-1860, Center of Military History of the United States Army, 2014

Stuart Reid, The Texan Army 1835-1846, Osprey Publishing, 2003

Gregory J.W. Urwin, The United States Cavalry: An Illustrated History 1775-1945, Blandford Press, 1983

Gregory J.W. Urwin, The United States Infantry: An Illustrated History 1775-1918, Blandford Press, 1988

About the Author

Gabriele Esposito is a military historian who works as a freelance author and researcher for some of the most important publishing houses in the military history sector. In particular, he is an expert specializing in uniformology: his interests and expertise range from the ancient civilizations to modern post-colonial conflicts. During recent years he has conducted and published several researches on the military history of the Latin American countries, with special attention on the War of the Triple Alliance and the War of the Pacific. He is among the leading experts on the military history of the Italian Wars of Unification and the Spanish Carlist Wars. His books and essays are published on a regular basis by Osprey Publishing, Pen & Sword, Winged Hussar Publishing and Libreria Editrice Goriziana; he is also the author of numerous military history articles appearing in specialized magazines like Ancient Warfare Magazine, Medieval Warfare Magazine, The Armourer, History of War, Guerres et Histoire, Focus Storia and Focus Storia Wars.